Veterinary Office Practices

Second Edition

Vicki Judah

DELMAR
CENGAGE Learning

Australia • Brazil • Japan • Korea • Mexico • Singapore • Spain • United Kingdom • United States

**Veterinary Office Practices,
Second edition**
Vicki Judah

Vice President, Editorial: Dave Garza

Director of Learning Solutions:
Matthew Kane

Senior Acquisitions Editor: Sherry Dickinson

Managing Editor: Marah Bellegarde

Editorial Assistant: Scott Royael

Vice President, Marketing: Jennifer
Ann Baker

Marketing Director: Deborah Yarnell

Marketing Manager: Erin Brennan

Senior Production Director: Wendy Troeger

Senior Content Project Manager:
Betsy Hough

Senior Art Director: David Arsenault

Cover Image Credits:

Ergonomic keyboard: © Peng Wu/iStockphoto

Dog at the veterinarian:
© Fuse/Getty Images

Reviewing medical files: ©spxChrome/
iStockphoto

Med Tech on phone: ©DenGuy/iStockphoto

Background image (dog biscuits in jar):
© Paige Worth, MWDesign

For product information and technology assistance, contact us at
Cengage Learning Customer & Sales Support, 1-800-354-9706
For permission to use material from this text or product,
submit all requests online at **www.cengage.com/permissions**.
Further permissions questions can be e-mailed to
permissionrequest@cengage.com

Example: Microsoft ® is a registered trademark of the Microsoft Corporation.

Library of Congress Control Number: 2011927909
ISBN-13: 978-1-111-13900-1
ISBN-10: 1-111-13900-8

Delmar
5 Maxwell Drive
Clifton Park, NY 12065-2919
USA

Cengage Learning is a leading provider of customized learning solutions with office locations around the globe, including Singapore, the United Kingdom, Australia, Mexico, Brazil, and Japan. Locate your local office at:
international.cengage.com/region

Cengage Learning products are represented in Canada by Nelson Education, Ltd.

To learn more about Delmar, visit **www.cengage.com/delmar**
Purchase any of our products at your local college store or at our preferred online store **www.cengagebrain.com**

Printed in the United States of America
1 2 3 4 5 6 7 15 14 13 12 11

For Don and Dorene Curry

Who have the strength and beauty of the great
forests and for being the most steadfast anchors
in the Sea of Cortez

Thank you

Contents

CHAPTER 6   Interacting with Clients   105

CHAPTER 7   Stress   141

CHAPTER 8   Ethics   153

Preface

Veterinary Office Practice is a required course for all students enrolled in an AVMA (American Veterinary Medical Association) accredited program leading to a degree in veterinary technology. It is also taught as core curriculum designed for veterinary assistants and veterinary receptionists. This course covers the business and professional aspects of a veterinary practice, including ethical and legal considerations, client communications, scheduling, record keeping, and stress management. The book provides an overview of the duties and responsibilities of the veterinary team members needed to provide the most professional care not only for the clients who bring their companion animals into the practice but also for their animal companions, the patients.

Organization of the Text

The second edition of *Veterinary Office Practices* has been expanded to provide a more in-depth discussion of the defined roles, duties, and responsibilities of the veterinary team members. There is a greater emphasis on cross-training and interpersonal communication skills. There is more information on health and safety including the potential risk of zoonotic diseases and interpreting basic animal behaviors, both essential to the welfare and safety of all veterinary staff and members of the public. Multiple applications of computers in the practice are explained with many new examples and in greater detail. Sample forms more accurately reflect the modern veterinary facility: why they are required and how they are used to benefit both the practice and the client, and in some instances, the legal requirements within the practice. Many of the photographs and illustrations are new to more accurately reflect the modern practice and material discussed in the various chapters. This book has been designed for those students seeking employment in a veterinary practice and acquiring the skills needed to become a valuable team member; it is about the belief that success, both personal and professional, comes from knowledge, teamwork, empathy, and compassion.

Veterinary Office Practices is divided into three parts that address different practice environments, the veterinary team members, and issues that complement the standard curriculum of a course in veterinary office practices.

Part I addresses the practice environment, specific members of the veterinary team, and their roles. Various types of practices are presented along with health and safety concerns for all employees. Duties of staff members, such as keeping/filing medical records and making appointments, and the importance of the computer systems are also discussed. Every practice has a variety of forms and brochures which the student will need to become familiar with and many of those are presented here. Part II focuses on the individual and the methods to develop personal communication skills, interactions with clients on the telephone and in person, and how to manage, prevent, and cope with stress. It also includes the ethical and legal responsibilities, the duty to society, confidentiality, and the veterinarian–client–patient relationship. Part III addresses financial concerns such as payroll accounting, taxes, and unemployment insurance, but the "business of the business" has not been addressed in the second edition. These issues are not normally within the realm or responsibility of the front-office staff and are, in most practices, dealt with by the practice owner, an accountant, or a professionally qualified veterinary practice manager.

Features

Veterinary Office Practices has been organized to provide current information and to facilitate study and review with features such as:

- *Chapter Objectives.* Students can use the list of objectives that accompanies each chapter to help them understand what they should learn and where to focus their attention.
- *Key Terms.* The most important terms in each chapter are identified in a list at the outset and then identified in bold type when they first appear in text.
- *Chapter Summaries.* A brief synopsis is included at the end of each chapter, representing the main points and helping students to tie the material together.
- *Review Questions.* Each chapter includes a list of review questions that test students' comprehension of the material, helping them to determine how well they understand what they read.
- *Scenarios.* Students learn better when they can apply concepts explained in a text to a real-world setting. With this in mind, scenarios exhibiting the practical application of instructions and information supplied in each chapter have been included to aid students' comprehension.
- *Online Resources.* Each chapter also includes suggested online resources, with search words for specific topics addressed in the text and related content.

New to this Edition

- Information on zoonotic disease potential
- Duties of all staff members redefined and expanded to include all professional organizations and requirements
- Discussion of the orientation period and what could be expected
- Discussion on how to approach and handle substance abuse
- The use of more appropriate terminology in describing tasks
- Expanded discussion on the use and application of computers in the practice
- Photographs and diagrams to reflect the modern veterinary facility

- A more in-depth discussion of the euthanasia process and coping with grief
- New material has been added to reflect the issue of veterinary ethics
- The addition of situational ethics regarding practicing veterinary medicine without a license and the duty to the veterinary profession
- A discussion of treating and possession of wildlife
- Information on specific duties of the front-office management team
- Additional and updated online resources at the end of each chapter

Instructor's Manual

Also offered with this book is an *Instructor's Manual to Accompany Veterinary Office Practices*, a chapter-by-chapter companion with classroom aids, featuring:

- answers to scenario study questions
- answers to review questions
- classroom exercises
- test bank questions and answers

About the Author

Vicki Judah is a licensed veterinary technician (Ret.) with 45 years of experience in animal care with a special interest in exotic animal medicine. Vicki is retired from the Jordan Applied Technology Center in Sandy, Utah. She developed the curriculum for a program in Companion Animal Science that was approved by the State Board of Education with endorsements to teach Animal Care and Science, Veterinary Assisting and Technology, Equine Science, Level 1 & 2, Agricultural/Livestock Care and Management, and Aquatics, Level 1 & 2. Vicki is now dedicated to developing educational material, writing and is involved in animal health, education, and rescue projects in Baja, Mexico.

Acknowledgments

Thanks are due to the many people who contributed to the revision of this text: Kathy Nuttall, Jordan Applied Technology Center, for her contribution of many new photographs and access to her teaching notes and material; Martin G. Orr, DVM, and his front-office staff, Draper, Utah; Eric L. Malaker, DVM, and the staff of Rochelle Veterinary Hospital, Rochelle, Illinois; Todd Bessendorfer, DVM, and his outstanding front-office team at Riverton Veterinary Hospital, Riverton, Utah. Also to Steve Wilton, SV Ocean Blue II, "the computer guy" who restored the files and countless other technical concerns encountered during this project including reviewing the chapter on computers, to Diana Ferraro for her help in loading a photo disk from multiple files, to Connie McWilliam-Schultz for her generous use of Internet time and friendship, and to Chris Bordato of LawDepot.com and of course to Benjamin Penner and Marah Bellegarde… My heartfelt thanks to all of you.

The author and Delmar wish to thank the following reviewers for their time and content expertise:

Steven W. Forney, MBA

Program Director of Skills and Trades Training

Coastal Carolina Community College

Jacksonville, NC

Sheryl Keeley, CVT

Veterinary Technology Program

Northwestern Connecticut Community College

Winstead, CT

Susan Kopp, DVM

Professor, Veterinary Technology Program

LaGuardia Community College

Long Island City, NY

Stuart Porter, VMD

Program Director, Professor of Veterinary Technology

Blue Ridge Community College

Weyers Cave, VA

Connie Maedke, CVT

Instructor, Veterinary Technology Program

Moraine Park Technical College

Fond du Lac, WI 54936

Rose Manduca, DVM

Professor, Veterinary Technology Program

Miami Dade College, School of Health Sciences

Miami, FL

Heather Wipijewski, CVT

Instructor, Veterinary Technician Program

Madison Area Technical College

Madison, WI

PART 1

The Veterinary Office

CHAPTER 1

Introduction to Veterinary Practice

OBJECTIVES

When you complete this chapter, you should be able to:
- identify and describe the roles and duties of various staff members within a veterinary practice and the professional organizations associated with them
- describe different types of veterinary care settings
- explain what cross-training is and how it can benefit both the individual and the practice
- explain the purpose and importance of an office procedures manual and a personnel manual

KEY TERMS

veterinarian
American Veterinary Medical Association (AVMA)
veterinary technician
National Association of Veterinary Technicians in America (NAVTA)
veterinary assistant
receptionist
practice manager
cross-training
personnel manual
procedures manual

Introduction

Practice management is a critical factor in the success or failure of any veterinary practice. Sound management techniques are necessary in every veterinary practice, whether it is a specialty referral hospital, a companion animal practice with multiple veterinarians, or a livestock practice with one veterinarian who makes farm calls.

When you think of a veterinary practice, you usually think first of the veterinary health care team who provide medical care to small animals: dogs, cats, and sometimes birds and exotics, or perhaps the familiar "vet truck" on country roads tending to horses and large animals on the farm. Veterinary care may be delivered in a clinic or hospital, a client's living room, in an outdoor enclosure, or even an open field. However, providing medical care is only one aspect of what is involved in a successful veterinary practice. There are many aspects of the practice that have to function with coordinated teamwork in order for veterinarians and their staff to be able to provide the professional care demanded of them (Figure 1-1).

Coordination and efficiency of this teamwork is basically a definition of veterinary practice management. All members of the veterinary health care team are involved in practice management issues, although "practice manager" may not be part of their title or job description.

Staffing the Practice

Who are the individuals involved in a veterinary practice? The first person who comes to mind is the **veterinarian**, a doctor who specializes in animal medicine. Veterinarians may be general practitioners who see most species of

FIGURE 1-1 Working in a busy veterinary office requires teamwork and includes more than assisting the veterinarian with medical care.

Photograph by Kathy Nuttall, courtesy of Riverton Veterinary Hospital, Riverton, UT animals, large and small. Some may specialize in internal medicine or orthopedic surgery, for example, while others prefer to work with zoos, wildlife parks, and aquaria. He or she is responsible for diagnosing illnesses, vaccinating against diseases, prescribing medications, performing surgery, and treating wounds. They may also be involved in preventative medicine and active in issues concerning public health. Veterinarians are also employed in industry and research.

To become a veterinarian, a person must earn a degree (Doctor of Veterinary Medicine (DVM), or University of Pennsylvania grants a VMD) from an accredited college of veterinary medicine associated with the campus of a state university. Acceptance into a school of veterinary medicine can be a challenge and all applicants must first have a four-year degree in a specific pre-veterinary curriculum or a science degree in a related field. Upon graduation, the veterinarian must pass a state examination in order to obtain a license to practice. Continuing education is required for veterinarians to maintain their licenses. Many veterinarians are members of the **American Veterinary Medical Association** (AVMA), a professional association dedicated to advancing the field of veterinary medicine and all of its related aspects. Many veterinarians also belong to their state association, for example, OVMA, Oregon Veterinary Medical Association.

Most veterinary practices employ at least one certified, registered, or licensed **veterinary technician** who works very closely with the veterinarian in providing medical care. The veterinary technician is a trained professional, much like a nurse in human medicine. Veterinary technicians must also attend an accredited college or university to attain a degree either as a veterinary technician (two years) or as a veterinary technologist (four years). They must also pass a state board examination to practice and to use the terms *recognized* and *implied* by obtaining a formal education and a successful examination result. The terms *certified, registered,* and *licensed* are not interchangeable but depend on the state laws governing veterinary technicians and which title is used by that particular state. Veterinary technicians and veterinary technologists are required to attend continuing education classes.

The technician is responsible for many aspects of the practice, including client education, advising on nutrition, and treating patients within the hospital. Veterinary technicians also fill prescriptions, administer medications, and give injections. They take medical histories and perform patient examinations. Additionally, they perform diagnostic tests with blood samples, analyze fecal and urine specimens, and calibrate the equipment used in the laboratory (Figure 1-2). Veterinary technicians also assist in surgery and/or deliver and monitor anesthesia. They also are responsible for taking and developing patient radiographs. All veterinary technicians are involved to some degree in practice management, depending on the size and organization of the clinic or hospital. They also have their own organization, the **National Association of Veterinary Technicians in America** (NAVTA), dedicated to fostering high standards of veterinary care, promoting the health care team, and advancing the veterinary technology profession. As with the AVMA state associations for veterinarians, there are also state associations for veterinary technicians.

Most practices also employ **veterinary assistants** who assist veterinary technicians and veterinarians. A veterinary assistant may be asked to restrain

Photograph by Kathy Nuttall, courtesy of Riverton Animal Hospital, Riverton, UT

FIGURE 1-2   One of the duties of a credentialed veterinary technician is to perform various laboratory tests.

an animal while the veterinarian or technician performs a physical examination or draws a blood sample, or to assist in the nursing care of hospitalized animals. Often, too, veterinary assistants are responsible for the cleanliness, exercise, and feeding of hospitalized patients. An important part of their job may be to assist in surgery by opening surgical packs, gowning and gloving the surgeon, and assisting the technician or anesthetist. Their role is important in cleaning surgical instruments, sterilizing and preparing the surgical packs. Additionally, veterinary assistants answer the telephones, schedule appointments, and help to keep the practice clean. Veterinary assistants, generally attend a school offering a veterinary assistant program, or enroll in a distance learning opportunity which would provide a certificate of completion. Most often, they are trained on the job by a veterinarian or a veterinary technician. Veterinary assistants are not eligible for licensing or certification at this time, and most states regulate the duties a veterinary assistant may or may not perform; for example, it may be unethical or illegal for a veterinary assistant to deliver anesthesia or perform certain laboratory tests. Currently, there is no national organization dedicated to bringing recognition and respect to the veterinary assistants' career; however, many state chapters of NAVTA welcome and support veterinary assistants as members.

Technicians and assistants provide a great deal of patient care and treatment but it must always be directed by and under the supervision of a veterinarian. *Only the veterinarian can diagnose medical conditions, prescribe medications, and perform surgery.*

In addition to those individuals delivering medical care, a veterinary clinic or hospital employs a **receptionist** who answers the phone, greets clients as they enter, coordinates the scheduling of patient appointments, and otherwise oversees the workings of the front desk area. There may be one designated head receptionist and other members of staff who are able to fill in as needed. Receptionists have a very important role in the professionalism and perception of the practice. Often, it is the receptionist who has the first contact with a client either on the telephone or in person. First impressions are lasting impressions and the receptionist is the most visible "first impression." The importance of being friendly, open, and welcoming cannot be overstated.

Many veterinary practices also have on staff a CVPM, a Certified **Veterinary Practice Manager**, a person whose specific education and experience are in the area of veterinary practice management. The educational requirements to become a CVPM are different from those on the medical team as their focus is business associated, dealing with human resources, finance, marketing, legal compliance, and ethics within the practice. They are responsible for employment issues, hiring and firing of staff, and employee evaluations and payroll. Anyone who desires to become a CVPM must, at first, have a minimum of three years working directly with practice management. An applicant must also obtain 18 semester hours (college credits) in courses related to business and management, and an additional 48 hours of continuing education directly related to management issues. It is necessary to pass an examination for certification. In many cases, the practice manager is also a credentialed veterinary technician. Certification as a CVPM is an excellent opportunity for advancement not only for the credentialed technician, but receptionists and veterinary assistants as well. Their professional organization is the Veterinary Hospital Managers Association.

The Role of the Veterinary Staff in Practice Management

As a veterinary technician, receptionist, or veterinary assistant, your role in the management of the office will vary dramatically from practice to practice. You may have no occasion to write an invoice or complete a transaction with a client. You may never be asked to keep track of the office supplies used in your practice. In contrast, part of your duties may be to routinely participate in the office activities of the practice where you work. You may be assigned to cover the front desk, answer the telephone, and be expected to file medical records that have been appropriately processed. Many students express the idea that they want to work with animals because they don't want to work with people. This is an impossible approach to having a successful career in the veterinary field because interaction with any animal also involves working with people and working to benefit the practice as a whole. It shouldn't matter how mundane the task may seem to be at the time, it is important to remember that everyone contributes to the success of the practice and the welfare of its patients. (Figure 1-3).

Orientation

The first few days on any new job can seem overwhelming with the amount of information there is to learn and assimilate. One thing to keep in mind is that

FIGURE 1-3 Some tasks may seem mundane, but keeping files in order is essential to the smooth running of the office.

there are usually many applicants for any position in a veterinary clinic. When you begin to feel overwhelmed remember that you have been selected to fill the position because the employer recognized your abilities and enthusiasm, your ability to get along well with others, your confidence, cheerful attitude, and willingness to learn. No one will expect you to know everything and it is important not to project the idea that you do but rather that you are ready to learn.

You will probably be assigned to "shadow" an experience member of staff who will ease you into the daily routine, demonstrate the duties and tasks expected of you, and how best to approach them. It is a good idea to carry a small notebook and write down things you don't understand or other questions which may arise during the day.

After a general tour of the facility and a chance to meet the other staff on duty, one of the first things to approach in becoming acquainted with a veterinary practice is to learn how time usually flows throughout the day and what the routine for a typical day is like and the services that are offered. You must become comfortable with the important details of time management within the veterinary practice and yet be aware that "routine" can quickly become chaos if the staff are not flexible when the unexpected occurs: emergency patients, other staff missing work, a doctor delayed because of surgery, or a computer "crash."

To help you get acquainted with the daily flow of activity, your mentor and now team member will provide many of the answers for questions you may have. Now is not the time to interrupt with questions regarding personal issues, such as when is lunch hour or what time do I take my break? These issues and other terms of employment would have been addressed at your time of hire or subsequently explained to you by the CVPM. Under no circumstances should you discuss salary, hourly pay, or benefits with any other staff member.

Introducing a new staff member to the routines of the clinic is usually based on sequential information: the what, where, why, and how of things so that they may be more easily learned and retained. During this period of orientation, you will be provided with a lot of information and the answers to many of your questions:

- The hours the practice is open to see clients and patients
- The priority in which patients are seen: emergency, scheduled appointments, and walk-in clients
- Telephone protocol, exactly how a staff member should answer the telephone. Listen carefully to how the telephone is answered and how callers are responded to.
- Are surgical and dental procedures scheduled for certain days or is a certain amount of time blocked off every day for these procedures?
- The method of the filing system or systems will be explained. You may in fact be asked to do some filing, either pulling files for the day's appointments or dealing with files which have been completed and need to be re-filed.
- The telephone system will be explained, the policy regarding putting callers on hold and how to transfer incoming calls and taking messages
- The use of the computer system will be introduced to you as well as the printer and fax machine
- Most veterinary clinics and hospitals have a retail section—a selection of food and products which are provided as an additional service to clients. Take the opportunity to explore this area of reception and become familiar with the products available.
- During these sessions, there will undoubtedly be incoming calls, client visits, and retail sales. Watch and listen closely regarding how these situations are handled.
- Step up and do the obvious: perhaps the magazines need to be straightened up, or there is a little bit of spot cleaning that needs to be done but above all, greet everyone who comes in with a welcoming smile. *You are now a member of the team!*

As the day progresses you will learn more about the scheduling of patient appointments, how they are scheduled, and who schedules them. It is unlikely that only the receptionist makes appointments, so you will need a clear understanding of how appointments are scheduled. Consider the following questions:

- If a client asks for a particular veterinarian in a multidoctor practice, how is it determined when the doctor is available to see patients and avoid a conflict with surgery or being unavailable for a particular day?
- What should the client be advised of in preparing an animal for surgery (i.e., time to arrive, withholding food from the patient for an appropriate time, and an approximate time a member of staff

will telephone regarding the surgery and advise when the patient may go home)?

- Exactly what paperwork is necessary to admit a patient? Which forms and releases must be completed and signed by the client?
- What are the procedures for discharging a patient? In some practices, every surgery is discharged by the veterinarian or a technician after a consultation with the owner. Sometimes the receptionist or veterinary assistant will be given discharge instructions to relay to the client and be responsible for the discharge.
- How long are the time slots for appointments (i.e., 10 minutes, 15 minutes, 20 minutes, 30 minutes)? (These openings will appear in the scheduling page.)
- How many time slots are set aside for a client bringing in more than one pet (i.e., to determine how many patients a client will be bringing in and schedule accordingly).
- What about "catch up" time during the day for the doctor to write medical records, return phone calls, perform procedures on in-house patients, and deal with the demands of a busy practice.
- Often, client/patient "re-check calls" are an important part of your duties. All details of the telephone conversation must be entered into the patient record even if there was no answer or a message has been left.
- If the veterinarian receives a call from a client, how should you advise the client regarding when the doctor may be available to return the call?
- Should the veterinarian be interrupted for a telephone call while she is with a client? It is important to determine what circumstances would apply and these are usually very specific situations. It is not normal protocol to interrupt a client visit unless specific instructions are given. These may be, for example, a family emergency, or a consultation with another veterinarian on a critical case. The veterinarian will usually advise the staff of any important calls he is expecting. Unless directed otherwise, the caller should be politely requested to a leave a message for the veterinarian who will return the call at his/her earliest opportunity.

Cross-Training

Cross-training is a simple, but powerful technique to enhance the value and versatility of every member of the veterinary health care team. Cross-training provides the skills necessary for you not only to do your own job, but to learn other staff members' duties so that you can perform their jobs if needed. This technique increases the flexibility of each staff member. It also increases the comfort level of staff members if one person gets tied up doing a particular task, is on vacation, or is called out of the office unexpectedly. Cross-training also makes your own job more interesting by providing variety and learning new skills. All of this will come in due time, when you gain experience in your own job and are comfortable with your required duties (Figure 1-4). Most veterinary practices find that having staff members

Photograph by Kathy Nuttall, courtesy of Riverton Veterinary Hospital, Riverton, UT

FIGURE 1-4   Cross-training is a benefit to all the staff. Here, a technician explains the benefits of a new drug to the receptionist who will be better prepared to answer a client's questions.

qualified to perform multiple tasks increases everyone's efficiency, reduces stress, and boosts personal morale.

Office Policies and Procedures

Most veterinary practices have a document that outlines the policies for the practice. It is usually referred to as the **personnel manual**. The personnel manual addresses issues such as attendance, dress code/appropriate attire, vacation, sick leave, codes of conduct, and multiple other issues which need to be clearly understood and adhered to by all members of the staff. These policies outline the employer's expectations and provide a basis of consistency in how practice issues are handled.

There is also a **procedures manual**, a separate document, or manual, from the personnel manual. The procedures manual contains the "how to" information about the daily activities in the practice. Examples of the information covered in a procedures manual include:

- opening and closing procedures for the facility
- how patients will be **admitted** and discharged
- what to do in case of fire, emergency, attempted robbery, client or staff injury, and dealing with an extremely angry or difficult client.
- how the phone is to be answered (every practice has its own preference and style)

- how receipting and invoicing should be handled
- information about accepting credit cards and handling client requests for credit
- the basics about the computer system, including end-of-day procedures, and closeout, or the shutting down of the system

The procedures manual outlines, in detail, the day-to-day workings of the practice or hospital. You should become very familiar with the practice's policies and procedures, to know where to look in the manuals for clarification, and to be aware of any additions or changes in the manual. The personnel policies and procedures manual provide a foundation for office management with specific standards that every one should follow, as all of these are very important to the well-being of everyone and the success of the practice. You may be well be asked to "sign off" on certain sections of the manuals, that is, your signature is a recorded (and legal) recognition that you have read and understood what is required of you.

SUMMARY

In addition to the veterinarian, veterinary practices also employ veterinary technicians, veterinary assistants, practice managers, and receptionists, each of whom plays a vital role in the day-to-day functioning of the office. Veterinarians and veterinary technicians are primarily health care providers. Other important members of the veterinary team are the veterinary assistants and receptionists. There is also usually a practice manager whose focus is on the administrative end of the business. Veterinary assistants, however, fall somewhere in between, aiding the medical practitioners and fulfilling a variety of office duties.

Veterinary care is provided in a variety of settings, including individual and group practices, veterinary specialty hospitals, on farms and in open fields, in zoos, aquaria, and research facilities. Regardless of the setting, as a member of the veterinary team you need to understand what is expected of you and be able to perform as needed. Be aware of how the practice does business, and become familiar with the policy and the procedures manuals.

SCENARIO

CJ arrived for her first day on the job as a veterinary assistant. She arrived a few minutes early, laid her coat on a bench in the waiting room, and nervously took a seat next to the receptionist, awaiting direction and instructions.

They sat and talked about the practice for a few minutes, and then two people came in with their pets. CJ watched the receptionist as she greeted the patients and took their information. The phone began to ring and as the receptionist was still busy with the clients, CJ decided to answer the phone:

"Um, hello?"

"Is this Clear Creek, what number is this? I am trying to reach the vet's office?"

"Yes this is Clear Creek. Can I help you?"

"Yes, I need an appointment today, the earlier the better. My dog refuses to get up. She was fine when my son took her out earlier."

"Hmmm… She probably just twisted something when she was out running and playing. Dogs do that a lot. I bet she'll be fine in a few days."

"I'd really need to have her seen. I need an appointment, the earliest available."

"I don't know. I guess you could bring her by."

"What time?"

"I don't really know, I just started here and I'm not the receptionist. Could you call back later? Maybe in a half hour or so?"

"No, I can't. I need an appointment!"

"Um… I'll see if I can put you on hold."

CJ thought she had put the client on hold but had in fact disconnected her. When the receptionist asked CJ who had called, she had no idea. Within the hour a very upset and angry client arrived with her golden retriever. (As later revealed by a radiograph, the dog had accidentally swallowed a bright green, and very slimy tennis ball.)

- What could CJ have done to better handle the situation?
- What could the receptionist have done to make things more comfortable for CJ and avoid the entire situation?
- What should CJ have been doing the morning of her first day on the job?
- What kind of preparation might have made this situation better?

REVIEW

Indicate whether statements 1–5 are true or false and if false, what is needed to make the statement true.

1. A veterinary assistant may be responsible for performing diagnostic analysis of blood samples.
2. The veterinary assistant is solely responsible for discharging patients after surgery.
3. A veterinary technician gives medication to animal patients and performs laboratory tests.
4. A veterinary assistant may be responsible for administering anesthesia or other medications during surgery.

5. Appropriate dress requirements for the practice in which you work will be outlined in the procedures manual.

6. List five policies of the practice that you will need to know if you will be scheduling appointments.

ONLINE RESOURCES

American Veterinary Medical Association (AVMA)

The official AVMA site features news articles, links to professional publications, educational resources, government regulations, notifications of disease concerns, a veterinary career center, and member discussions.

<http://www.avma.org>

National Association of Veterinary Technicians in America (NAVTA)

The NAVTA Web site features information on promoting veterinary technicians, career building and credentialing information, listings of state representatives and student chapters, the veterinary technician code of ethics, and links to related sites.

<http://www.navta.net>

Employer-Employee.com

This Web site focuses on issues faced in many office settings, including issues of policy and procedure, management, team building, and improving lines of communication.

http://www.employer-employee.com/

Veterinary Hospital Managers Association

This Web site provides greater detail about the organization, history, and credentialing requirements.

CHAPTER 2

Care and Maintenance of the Veterinary Practice Facility

OBJECTIVES

When you complete this chapter, you should be able to:

- explain the importance of maintaining a clean, safe, and organized veterinary facility, and describe what steps must be taken to do so
- explain what you should do when you find that a piece of equipment or some aspect of the facility needs to be repaired
- identify common safety hazards
- identify common sources of accidental injuries and what you can do to protect yourself
- explain general techniques for avoiding infection and the spread of disease among humans and animals in the practice
- explain what OSHA is and describe some of the general guidelines that must be followed in the practice

KEY TERMS

Occupational Safety and Health
  Administration (OSHA)
zoonotic disease
clean as you go
footbath
giardia
toxoplasmosis
radiology
dosimitry badge
induced

inhalent
sharps container
pathogens
right-to-know station
material safety data sheets (MSDS)
capital equipment
medical supplies
pharmaceuticals
inventory
want list

Introduction

The general maintenance of a veterinary practice could be a full-time job. There are many aspects that are very important, not only for the way a practice functions, but for the health and safety of the staff, clients, and patients. The cleanliness of the facility is critical, not only to ensure delivery of high-quality care to patients, but also to communicate to clients the quality of service delivered. Plans must be in place to deal with the unexpected breakdown of an essential piece of equipment, whether medical such as an anesthesia machine or nonmedical such as the air-conditioning unit or the furnace.

All veterinary practices are required to comply with the Occupational Safety and Health Act of 1970, regulated by the **Occupational Safety and Health Administration** (OSHA), which provides job safety and health protection for workers by promoting safe and healthy working conditions.

Most veterinary practices have identified specific strategies to help their employees prevent injuries. You may be required to read and understand a **safety manual** in addition to the personnel manual and the procedures manual. There are also safety precautions that must be taken to help minimize hazards to clients visiting the practice. Preventing the spread of disease within a veterinary facility from one patient to another is an important aspect of hospital care and maintenance.

You must also be aware that some diseases have **zoonotic** potential. A zoonotic disease is one that can be transmitted directly from animals to humans. The causative agent may be bacteria, virus, fungus, protozoan, or a parasite. Common routes of transmission include inhalation, direct contact, exposure to urine and fecal material, contaminated food/water bowls, handling dead or diseased animals and their body fluids and tissues. Methods of transmission can be as variable as each disease causing organism. Table 2-1 lists some of the common zoonotic disease and their routes of transmission.

Washing your hands between patients and the items associated with their care is extremely important in helping to prevent the spread of disease and decreasing zoonotic potential (Figure 2-1). There may also be situations in which disposable gloves should be worn, for example when cleaning cat litter trays and handling other animal waste.

Maintaining a Safe Facility

Housekeeping and General Cleaning

If it smells clean, then it must be clean.

These words address one of the most important aspects of the general care and maintenance of the veterinary practice: odor control. It is impossible to achieve adequate odor control in any veterinary facility without making cleanliness a priority. A clinic may look clean, but if there is a constant odor, either

TABLE 2-1   Safe Practices

- Exotics should not be allowed to roam freely. This especially applies to reptiles.
- Reptiles should be provided with dedicated tubs for soaking, not placed in the bathtub or sink.
- No animal should be in or near food preparation areas.
- Animals should only be fed from dishes reserved exclusively for them; never from hands or allowed to take food from a human mouth, to lick dinner plates and other utensils.
- All animal food and water dishes should be disinfected and washed separately, not placed in a dishwasher.
- Animal food and water bowls should not be exchanged between species (e.g., the snake bowl should not be used to provide water for the family dog).
- All soiled bedding and fecal material should be handled with latex gloves and disposed of in tied plastic bags. It should not be used as garden mulch!
- Hands should ALWAYS be thoroughly washed with soap and water after handling any animal, animal equipment, and enclosure contact, even if wearing disposable latex gloves.
- Latex gloves should be pulled off so they are inside-out, tied at the wrist, and disposed of in a plastic bag.
- Hands should be washed between handling other animals.
- In cleaning cages that are very dirty, a disposable face mask should be worn.
- When handling a diseased animal, protective eye wear should also be worn.

FIGURE 2-1 Washing your hands is one of the simplest methods of reducing disease transmission.

© Delmar/Cengage Learning

musty (a soiled mop left in a bucket of dirty water) or it smells of animals, then general cleanliness is not receiving enough attention.

Many veterinary practices use professional cleaning services to maintain the facility. The service provides floor cleaning and polishing, washing down the walls and a window cleaning service. The schedule of a professional cleaning service will vary with the needs of the practice. A high-volume practice may have the cleaning service come in every night after the clinic is closed, while another practice may have professional cleaning done weekly or monthly. Whether or not your office uses a cleaning service, it is essential that every staff member contribute to the cleanliness of the practice.

It is everyone's responsibility to help keep the practice clean on a constant and consistent basis. For many veterinary assistants, the job description includes cleaning cages and feeding, watering, exercising, and bathing animals, all of which relate to keeping the practice clean. You will need to learn what cleaning products are used in the different areas of the facility. A chemical that works well on the floor and cuts grime in high traffic areas might actually cause harm to an animal if it were used inside a cage, a dog run, or on a stainless steel table in the examination room. If you are not sure whether a product is safe to use in a particular area, ask! Industrial cleaners can have very harmful effects on humans and animals if used improperly.

Have pride in the facility where you work, and contribute wherever you can to the overall housekeeping of the practice. You will probably have the opportunity to participate in activities and procedures in many different areas of the practice and you will become aware of the specific cleaning needs in different areas of the hospital, for example, the surgical room may require different and daily routines whereas the storage areas and supplies cupboard may be dusted on a weekly basis. Most cleanliness issues involve common sense. Make it standard practice to **"clean as you go."** This simply means putting things back where you found them when you are finished using them, and cleaning an area after a procedure. It is very important to clean the exam room after each client so that it is ready for the next person who enters the room (Figure 2-2). Every work station and exam room should have a bottle of disinfectant spray cleaner. These spray bottles should also be kept clean and filled with the approved product for the purpose. It is essential to clean up immediately when a patient soils the floor or a male dog "cocks his leg" and urinates on the wall or furniture in the waiting room.

Most veterinary practices want to minimize cleanup of animal waste which is not only a source of odor, but a source of spreading disease among animals. One way to accomplish this goal is to recognize that dogs need to be walked outside in an exercise area on a regular time schedule so that they may empty their bowels and bladders. It is a natural behavior of dogs not to soil where they eat and sleep and most of them have been housetrained and recognize that urinating and defecating inside is unacceptable. Be aware of their comfort by providing them with adequate exercise and elimination opportunities. You will enjoy the additional benefits of decreased cleanup in the kennel and cage area, decreased risk of offensive odors, and, most importantly, decreased risk of disease transmission within the veterinary facility. It is also

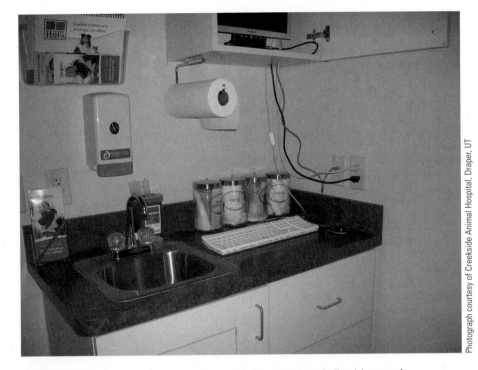

FIGURE 2-2   The exam room should be sparkling clean and all evidence of the previous patient should be removed and the trash emptied.

very important that the outside exercise and elimination area is kept clean, the feces collected, bagged, and disposed of correctly. Many facilities make use of a **footbath**, a small shallow tub which contains a disinfectant and helps to prevent possible transmission of disease from the outside area to inside the hospital. This is a good preventative measure to help stop the spread of **giardia**, a microscopic parasite which is transmitted in feces and through contaminated grass. Giardia is zoonotic and causes severe stomach pains and explosive, bloody diarrhea in humans.

When caring for cats, keeping litter pans clean is an important step in preventing undesirable odors and diseases. Most veterinary facilities have a morning and evening routine that involves scooping cat litter pans, changing the water, and feeding. Regardless of the routine established, litter boxes should be cleaned whenever they become soiled. Many feline patients often curl up or sit in their litter box in preference to other bedding provided. They seem to feel more protected, especially when they are in a totally exposed cage. Whenever a litter box is cleaned you should always wear disposable gloves (Figure 2-3). If the litter is particularly dusty, a non-clumping variety, sometimes wearing a paper mask is also appropriate. The gloves should be removed by pulling them off so that they are inside out, one inside the other, and disposed of in the designated waste container (Figure 2-4). Use a separate pair of gloves for each patient and always wash your hands thoroughly. *Pregnant women should not clean litter boxes because of the zoonotic potential of* **toxoplasmosis**, *a microscopic parasite. Exposure to toxoplasma gondii can cause birth defects.*

Photograph by Kathy Nuttall, courtesy of Riverton Veterinary Hospital

FIGURE 2-3 Keeping cages and litter trays clean is an important part of the job in a small animal practice. When cleaning cat litter trays, gloves should be worn.

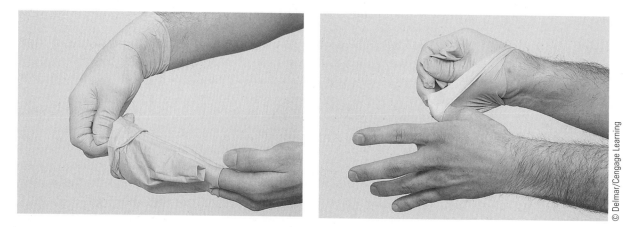

© Delmar/Cengage Learning

FIGURE 2-4 Gloves should be removed by pulling them off so that they are inside out and disposed of in the designated waste container.

If canned food is fed to cats, be sure to remove any uneaten portions and note which patient did not eat. Uneaten canned cat food can produce quite an unpleasant smell and it is the perfect place for the growth of mold and bacteria. Cleanliness also directly affects cats, which are very clean and particular animals. Cats will often refuse to use a litter pan when there is an accumulation of waste in their litter pans. If a cat is in the veterinary hospital because it is ill, you want to do all you can to encourage a healthy environment. Keeping a cat's cage clean is a good beginning.

Look around you throughout the day and ask yourself, "If my animal were staying here, would I be happy with the environment?" This is a good way to stay in touch with how comfortable and clean the animals are at any time.

Safety Hazards

In any veterinary practice, the potential is great to encounter a wide variety of safety hazards. Many of these hazards are unique to the veterinary profession. Animals may bite or scratch. There are needles, scalpel blades, and other sharp objects to dispose of properly. Chemicals, animal urine and feces, and the discharge from infected wounds can pose a health risk to humans if not handled properly. Within the veterinary facility, there may also be equipment that can cause injury.

Radiology, the area in which X-rays are taken, is contained in a separate room which is lead-lined. If you have occasion to assist in taking a radiograph, you will be provided with a lead apron, gloves, and a throat (thyroid) shield to further prevent exposure to X-rays. Each member of the staff who participates in taking radiographs will have a **dosimitry badge**. Dosimitry badges measure an individual's exposure to scatter radiation. They are easily clipped on to the front of a garment or the lead apron being worn. Periodically, these badges are replaced with new ones and the exposed badges are sent to a company which measures and reports radiation exposure for each individual. It is very important that you do not use a badge that belongs to another member of staff and equally important that you do not participate in taking a radiograph without this personal monitor. There will also be a bold sign on the door to the radiology room and a warning to pregnant women. Radiation can be very harmful to an unborn child. You should advise your supervisor if you know that you are pregnant or even think that you might be.

Lifting is another common source of injury. It is very important to know how to lift correctly whether it is a large dog or a bag of food. Always ask for help and do not attempt to lift heavy items or patients on your own.

Some of the medications used in a veterinary practice can also pose a variety of health risks. For example, a common antibiotic used in animals is amoxicillin, a relative of penicillin. It comes in a powdered form and it is mixed with water. Just inhaling some of the powder can cause a severe allergic reaction for some people who may be allergic to penicillin. If the hospital where you work treats cancer patients, you will be given instruction on how to avoid accidental exposure to potent chemotherapy drugs and you may be restricted entirely from the area when these drugs are being handled and used to treat a patient. There are also special safety regulations that now apply to the use of lasers during a surgical procedure.

On very rare occasions, there may a be leak in an anesthesia delivery system or an animal that is **induced** (the start of the anesthesia process) in one area is transferred to another delivery system and the first anesthesia system has been left open; for example, animals are usually induced and "prepped" for surgery in a separate area from the operating room. In the rush to get the animal "on the table," the first system is inadvertently left running. There are different types of **inhalent** anesthesia agents, those which are breathed in, but all are liquid and are turned into gas by the presence of oxygen. Each of these agents has a distinctive odor which you need to recognize. Your immediate action should be to turn on the ventilation system, ensure that the anesthesia delivery system is off, and report it immediately to the veterinary technician. It is unlikely that you will receive training in the delivery of anesthesia, but you should know which valves will turn off the supply of the anesthetic agent and the oxygen flow to the machine. Anesthesia delivery systems are precisely calibrated and this is not something you should guess about or "fiddle with"—ask for instruction before a situation like this arises. If you know that a particular system has not been used and you suspect a leak, contact your supervisor immediately.

Preventing Accidental Injury

The most common injuries in veterinary practices are bite wounds from animals. Dogs and cats usually come to mind first, but do not forget that ferrets, rabbits, birds, and reptiles can all inflict serious injuries as well. *All animals can and will bite.* Most bite injuries are preventable if animals are approached and handled correctly. It is important to learn the body postures of animals and this varies from species to species. If you do not have much experience handling and restraining animals in a hospital or clinic, you will be taught the various methods of restraint and handling. It is in everyone's interest and safety that patients are under control at all times. The sweet little dog that arrived in the morning may have transformed into a growling, snapping patient. Most of these aggressive behaviors are fear based: of being away from the owner, of being confined in a strange environment with multiple other animals and scents, and being approached, "cornered" in a cage. Never, *ever* believe an owner who says, "She won't bite, she's just scared." A fearful dog is much more likely to bite and with very little provocation. Handling animals safely takes practice (Table 2-2).

TABLE 2-2	Short list of Common Zoonotic Infectious Agents and Source Groups
Bacteria	**Source**
Salmonella	All animals
Aeromonas	Reptiles
Mycobacterium	Mammals, reptiles
C. psittaci	Birds
Listeria	Amphibians

continued

Table 2-2   (continued)

Tularemia	Mammals (especially rabbits)
Leptospirosis	Mammals
Campylobacter	Reptiles
Enterobacter	Reptiles
Streptobacillus	Mammals
Pasterurella	Mammals, birds
Colibacilla	Birds
Yersinia pestis	Mammals (rodents and fleas)
Tuberculosis	Mammals, birds
Staphylococcus	Mammals, birds
Viral	
LCM (lymphocytic choriomeningitis)	Mammals
Rabies	Mammals (rarely reported in birds of prey)
Monkey Pox	Potentially any mammal
Yeast	
Candida	Reptiles
Fungal	
Dermatophytes (ringworm)	Mammals
Cryptococcus	Birds
Protozoan	
Giardia	All species potential
Coccidia	All species potential

First, you will be taught how to handle the animals that are not stressed by being in a veterinary hospital. Each approach is different because each patient is different. You will be shown how to remove patients from their cages and return them safely after exercise or an exam, restrain them for physical examinations, take them outside to eliminate, or move them from one pen to another (Figures 2-5 and 2-6). It is important to learn the behaviors, postures, vocalizations, and facial expressions that animals use to communicate their attitudes. Constantly observe their behaviors as they are approached and handled by experienced staff so you can begin to interpret when an animal will allow you to do what needs to be done without a struggle, and when it would be best to call for an extra hand.

Most of the animals encountered in a veterinary practice are companion animals and are used to being handled. The first rule of restraint is "least is best" as many animals will react aggressively if the restraint technique is too harsh for the circumstances. If you are unsure of how to restrain a particular animal, ask for advice before you attempt to approach and restrain the animal. *It is critical that clients are not allowed to handle or restrain their animals while the animals are under the care of the veterinary practice.* If a client is injured by his own animal during the course of a procedure, the veterinary practice is held legally liable. As tempting as it may be to allow the owner to hold and reassure

FIGURE 2-5 The leash should be placed around the dog's neck before fully opening the cage door.

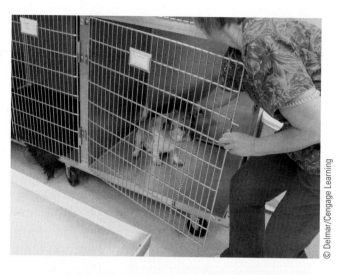

FIGURE 2-6 The patient should be allowed to walk out on its own from a cage at floor level.

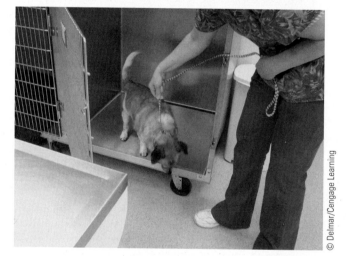

his or her frightened pet, it is simply not worth the risk of the client being injured and the hospital and staff being involved in a lawsuit for personal injury and negligence.

During cleanup procedures, be alert for needles or scalpel blades that have not been placed into the appropriate **sharps container**. These containers are made of heavy plastic with a lid that can be permanently sealed. When the container is full, the lid is sealed and the container is picked up by a service licensed to dispose of hazardous waste materials. When you find an unlabeled container, bring it to the attention of a senior member of staff immediately. By remaining alert to your surroundings, you usually can avoid circumstances that might otherwise lead to accidental injuries.

Controlling Infection and Spread of Disease

It is especially important, when working in a veterinary facility, to be aware of the risks of passing **pathogens**, any disease causing organism, from one animal to another. Most animals are admitted to a veterinary hospital because they are sick or injured. Their immune systems are often compromised from the stress of their illness or injury, and they are more susceptible to disease. The last thing a veterinarian, or any member of staff, wants is one sick animal serving as the source of illness for another. Animals in the hospital for surgical procedures experience more stress than usual—making them more susceptible to a pathogen if the veterinary staff are not vigilant.

For disease containment, like many other aspects of patient care, common sense will often help you make wise choices when you are handling patients. Animals suffering from infectious diseases should be kept in isolation ward to avoid infecting other patients. Potentially contagious patients should have their daily maintenance, like feeding and watering, tended to after the healthy animals. This decreases the risk of carrying a disease from the isolated patients to the other patients. Always wash your hands between handling each patient. Change into a clean smock (scrub top) whenever you have handled an ill animal or if the scrub top has been soiled with urine, stool, blood, pus, or other body fluids. Never wear the same scrub top for feeding, watering, or medicating animals that you wore while cleaning cages. A separate apron should also be used for bathing animals. It is also extremely important to never eat or drink anything in any of the animal treatment or laboratory areas. There is usually a staff lounge and you should only eat your lunch or sit to rest and have something to drink in the designated area.

You must learn and be comfortable with all procedures outlined by your veterinarian for reducing the risks of transmitting disease from one animal patient to another. If you come upon a situation with which you are not familiar, or if you are unsure how to proceed, you should always seek out a senior member of staff and ask for guidance.

There may be circumstances when a truly infectious case is hospitalized in the facility's isolation ward. You may or may not be authorized to handle that particular case. Every practice has its own requirements, and it is your responsibility to do all you can to keep the patients of the practice safe from unnecessary risk.

Osha Guidelines

All employers are required to abide by the Occupational Safety and Health Act of 1970. Commonly referred to as OSHA, it provides for safe and healthy working conditions for all employees. The guidelines were created by the U.S. Department of Labor. The employer is required to inform all employees of potential hazards in the workplace so that the employees understand how best to keep themselves safe. OSHA regulates many aspects of businesses in all industries in the United States. Examples of safety regulations that must be adhered to include the following, which require veterinary practices and hospitals to:

- keep the location clean, safe, and dry
- provide easy access to clearly marked, adequately sized, unobstructed exits
- ensure that oxygen tanks are secured and contained only in green cylinders
- follow strict safety guidelines when handling hazardous chemicals

- require the use of proper protection for eyes, face, skin, feet, respiratory system, hands, and any other body part in situations posing an environmental, chemical, situational, radiological, mechanical, or any other form of physical hazard
- maintain a readily accessible eye-wash station
- provide toilet facilities and clean water
- maintain operating fire extinguishers, sprinkler systems, and other fire extinguishing systems (requirements can vary by location, type of business, materials involved, etc.)
- include protective guards or covers for machinery
- keep electrical equipment free of hazards
- provide information to employees regarding all potential hazards of using equipment, hazardous materials, chemicals, and all materials contained within the facility.

This is only a small sampling of some general OSHA rules and guidelines. Your employer will have more information about the specific regulations that must be followed in the practice or hospital. Additional information about OSHA regulations is available online at <http://www.osha.gov>. OSHA can also be reached at 1-800-321-OSHA (6742), or via mail at U.S. Department of Labor, Occupational Safety & Health Administration, 200 Constitution Avenue, Washington, D.C. 20210.

While this information is provided and readily available elsewhere, employees should not contact OSHA directly regarding a situation or concern about their place of employment unless all other avenues have been explored to resolve the issue. Your immediate contact should be your supervisor who will take the proper steps to rectify the situation and resolve your concern. Filing a complaint with OSHA is only a last resort and it will be taken very seriously and investigated by OSHA.

Right-to-Know Station

Somewhere in the staff area of the practice, there will be a **right-to-know station** or a specific area set aside with some shelving where there is a notebook or binder containing the safety information that applies to that particular practice. Contained in the safety notebook will be **material safety data sheets** (MSDS), which are published by the manufacturers of the various products used by the practice. Read and understand this information to be aware of any special handling requirements of the products or any restrictions that may apply to the user of the products. This information varies for each product used by the practice.

The safety notebook should contain (1) evacuation plans in the event of an emergency; (2) the locations of gas and water valves for rapid shut off; (3) the locations of fire extinguishers; and (4) emergency telephone numbers for the police, the fire department, and companies that service the facility's heating, ventilation, and air-conditioning units. All chemicals, disinfectants, and other products that are used within the practice will be labeled by your employer with any appropriate hazardous considerations. In addition, personal safety equipment like safety glasses, latex gloves, and earplugs should be supplied by your employer.

Your employer, the veterinarian, or the designated safety officer within the practice will show you where the safety information and safety equipment are kept. It is your responsibility to read and understand the practice's safety manual, to know how to use the required personal safety equipment, and then to actually use it. It is also your responsibility to be comfortable reading MSDS for appropriate precautions. You must take the initiative to remain alert to potential hazards you encounter within the practice facility and then bring them to the attention of the appropriate individual. Safety is everyone's business.

In order to avoid losses from preventable injuries, most veterinary employers have developed strategies to help their workers anticipate potentially hazardous situations. Usually, they will supply you with a document that outlines the steps to take in the event that you witness or are involved in an injury at the practice. If you experience or witness an accident in which a staff member, client, or visitor sustains a personal injury, regardless of how minor it may seem, immediately report the situation to your supervisor or to the person in the practice who oversees safety issues. Failure to report an accident or injury can result in a violation of legal requirements and could delay the processing of appropriate insurance claims and benefits.

Office Equipment and Maintenance

Repairs and Troubleshooting

Part of helping a veterinary facility run smoothly involves keeping the equipment, both medical and nonmedical, in good working order. It is unlikely that you will be called upon to repair any of the equipment, but it is your responsibility to learn the appropriate use of any equipment or tools that are the property of the practice. If there is no formal training program in place to prepare you to use the equipment in the practice, seek out a supervisor who is able to demonstrate correct usage. You may also be asked to participate in regular, routine maintenance of the practice's equipment. Proper use and maintenance might include:

- replacing paper in printers, photocopiers, and fax machines
- replacing ink or toner cartridges in printers, photocopiers, and fax machines
- learning how to properly turn each piece of equipment on and off
- learning how to operate machinery and equipment you are not familiar with
- learning how to identify technical problems

Most veterinary practices are now computerized, and you will be working with a computer station and keyboard as part of your job. Although you are probably at least somewhat familiar with computers, you may not be skilled in using the specific veterinary programs used in the office. You will be given training in the use of these specific veterinary programs so you can use the entire systems correctly and effectively. (Computers and software are discussed in greater detail in Chapter 4.)

In addition to computers, you can expect to find and use many other (sometimes complex) pieces of office equipment. Sophisticated telephone systems for the veterinary practice allow multiple phone lines to come into a single reception area, and many systems have built-in voice mail and paging capacities. The practice where you work may use a telephone answering machine after hours. You may be asked to record the outgoing message and program the system to receive calls each day at the end of business hours. A fax machine, now common in veterinary offices, may be a stand-alone piece of equipment or it may also include a telephone. Your practice might also use a multifunction machine, incorporating a printer, fax machine, copier, and/or scanner.

You will be instructed on the policy and use of the receipts drawer or a cash drop box, a credit card processor, a postage scale, and a postage meter. The practice may use audiovisual equipment for client education or staff-training purposes. There may be a TV/VCR/DVD for use at the facility. No matter what equipment you find in the practice where you work, part of your responsibility is to become very familiar with how to use, care for, and manage it—and whom to contact if something unusual happens.

In addition to learning how to use various pieces of equipment to their best advantage, you need to learn to perform basic maintenance and troubleshooting. For instance, you should know how to load paper into the printers attached to the computers. The process is different if the printer is a bubble jet, laser, or dot-matrix type. Paper tapes in the credit card processing unit must also be replaced when they run out.

You may be asked to change the cartridge in a computer printer. If the practice has a photocopier, you may need to replenish toner or empty a full toner receptacle. You will want to become familiar with refilling the paper trays in the copy machine and the printer, as well as knowing where replacement paper is stored.

Although you may not be required to repair equipment in the practice, you will need to know the person in the practice to contact if there is a malfunction. Most likely, you will report a problem with front-office equipment, like the photocopier or the fax machine, to the office manager or receptionist. In the treatment and surgical areas of the practice, a senior veterinary technician may be responsible for the equipment. Be aware of the optimal temperature in different areas of the practice facility, so that you will notice any sudden changes. Be alert to hear the furnace making strange noises or an exhaust fan failing.

You should report any abnormalities that you notice immediately. Do not assume that the appropriate person already knows of the problem. To avoid causing further damage, do not use any equipment that is functioning abnormally or is sending a message that a part is not working. By ignoring a minor problem, it is likely to become a major one. You may be asked to contact the designated repair or service person, so you will want to be familiar with where a list of those support people is kept or who in the practice has access to that information.

Materials and Supplies

Veterinary facilities use many different types of materials and supplies. You would expect to find some of these in any office setting. Others, very specific to the veterinary profession, require detailed record-keeping techniques and many of the inventory records may be maintained through the use of a computer program.

The supplies that help office equipment operate will likely be found in the administrative area of the practice. Most often, the lead receptionist or the practice manager is in charge of keeping an acceptable **inventory** of those supplies. An up-to-date inventory keeps track of the quantity of each item on hand so there is always an adequate supply available without running out or creating a surplus. Supplies for the office area of the practice include copy paper, computer paper, printer cartridges (ink or toner), pens, pencils, markers, telephone message books, and anything else that is necessary for a smoothly functioning office. You will need to be familiar with the storage area for various office supplies in case you are working at the reception desk and find, for example, that the copy machine is out of toner. You should also know exactly who to talk to when you need to order more, and the requisition process, if applicable.

Another category of veterinary practice supplies includes the **capital equipment**—equipment that has a fairly long life expectancy and contributes to the income of the practice. Examples of capital equipment include in-house laboratory equipment and microscopes. Other capital equipment in a veterinary practice include anesthesia delivery systems, an X-ray machine, and surgical instruments.

A third category of supplies necessary for the smooth functioning of a veterinary practice is **medical supplies**. This category includes items like bandaging and suture material, intravenous fluids, needles, syringes, X-ray film, and chemicals that are used in the laboratory. These items, used in the everyday functioning of the practice, help the veterinarians and veterinary technicians do their jobs. Equally as important as medical supplies are the **pharmaceuticals**, the drugs and other medications used in the practice as well as those dispensed to clients from the practice pharmacy. You would not expect to find vaccines among the pharmaceuticals in the drug cabinet as these require refrigeration as do some other medications. You should become familiar with the layout of the drug cupboard as this may vary from practice to practice; some prefer to have the drugs organized by category, while others prefer alphabetical order (Figure 2-7). There will be no narcotics or scheduled drugs in the cupboard. These potentially dangerous drugs are kept locked in a separate cabinet.

Want Lists

Few things are more frustrating in any work situation than to be in the middle of a project or procedure and suddenly run out of needed supplies to finish the task at hand. In a medical setting, this scenario can have deadly implications. It is critical for all staff members to know the procedure used to avoid shortages of practice inventory, a complete and accurate list of supplies and quantities

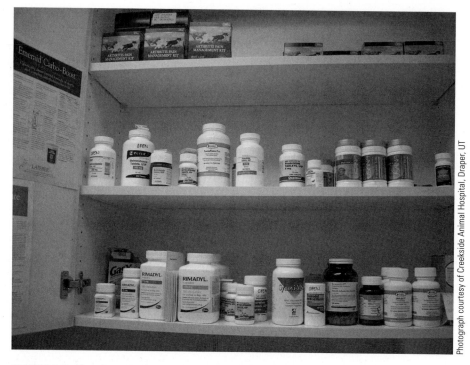

FIGURE 2-7 The drug cupboard should be well stocked and kept in an orderly manner.

on hand. Most practices have very specific staff expectations concerning keeping appropriate levels of products and supplies that should always be available.

One very useful tool is the **want list**—a list posted in a notebook, on a clipboard, or on a white board at a specific location in the practice for all to see and use. When you retrieve a particular medication or supply from the practice inventory and notice that the remaining quantity has dropped to a low level, it is time to enter that item on the want list. Most practices have a minimum quantity of commonly used products that is to remain on hand. At regular intervals, the person in charge of ordering practice materials retrieves the want list and uses it as a guide in purchasing items for a particular time period—whether ordering is done on a weekly, monthly, or quarterly basis.

Some practices use separate want lists for medical supplies and for office supplies. Breaking it down further, you may find a want list for the front-office area, another for the pharmacy, and yet another for the treatment/surgery area of the practice.

Inventory Management

Most practices and hospitals use computer systems to assist with purchasing decisions. If the inventory for the practice is entered into the computer system accurately and tied into the invoicing function of the practice program, any inventory items entered on an invoice will automatically be removed from the practice inventory. Periodically, the computer program can then be accessed to generate a report suggesting necessary purchases. When the ordered supplies arrive, each item and quantity is entered into the computer to update the inventory.

Usually, a veterinary practice has a single person responsible for placing orders for inventory items whether or not a computer is used in the purchasing process. That person may be the office manager, the practice manager, or the practice owner. A one-person system helps avoid duplication. It also avoids confusion—for example, one staff member may think that another person had placed an order—leaving the order undone. One of your jobs will be to record, in a timely fashion on the appropriate want list, any items that you notice are running low. You also may be asked to report shortages directly to the person involved with ordering.

Although you may not be involved directly in ordering, it is highly likely that you will be involved, in some way, in processing orders when they arrive. When an order is received, the first step is to be sure that there has been no damage to the package. If a shipment arrives damaged, you should follow whatever guidelines are appropriate for the particular carrier or delivery system. Step two is to open the package, find the packing slip or invoice, and carefully examine the contents to be sure that all items listed are present. This should be done prior to signing for the delivery, as your signature indicates that the shipment was received, with all appropriate contents and in acceptable condition. Report any shortages to the appropriate staff member so that the supplier can be notified. At this time, you should also record any listed product's expiration date. This procedure allows the person entering information about the order in the practice's computer to include these dates. It also provides a means of screening shipped products that have been *short-dated*—that is, products that will not be used up before their expiration date. This procedure gives the practice owner, or whoever handles inventory management, the opportunity to contact the company about an exchange as well as any shortages.

When you are confident that everything on the invoice is present and you have recorded the expiration dates of products when appropriate, put everything in its correct location in the practice facility. Certain drugs and lab tests require refrigeration or freezing. These should be checked in immediately and then stored in the refrigerator or freezer as necessary before moving on to the remainder of the delivery. Remember to *rotate* the inventory as you put things away. Rotating the inventory—moving products that are already on the shelves to the front and placing newer items behind—ensures that products are used by the practice or sold to the clients in the order in which they were received. Rotation cuts down on the risks that products will become outdated before they can be used or sold.

Occasionally, even with the most accurate computerized inventory records, you will be asked to assist with a *physical inventory*. This means counting every single inventory item in the practice and making a notation of the quantity on an inventory list. You may find that the list for office supplies is kept separately from the medical supplies list. The pharmaceuticals may be on their own list as well. Many practices conduct a physical inventory at the end of its fiscal year, which may not be the same as the calendar year. Some practices perform a physical inventory at the end of each quarter in an effort to have the inventory quantities in the computer remain as accurate as possible. This also helps to make each person aware of how great the inventory actually is.

SUMMARY

Cleanliness, safety, the containment of infectious disease, and the general maintenance of a healthy facility are responsibilities of all persons involved in a veterinary practice. Each member of the veterinary staff must be involved in identifying and correcting hazards and potential contamination, and be responsible for maintaining sanitary conditions for the animals in your care.

To maintain a safe working environment, all members of the veterinary staff should comply with OSHA regulations and use common sense in daily activities. If something is too heavy to lift alone, ask someone to help. If a piece of machinery poses a hazard, be sure to use care and caution while operating it, and do so only after being adequately trained.

Office equipment, materials, and supplies are also important to the smooth functioning of a veterinary practice, and the use of want lists helps with inventory management, ensuring that when something is needed, it is available and in optimal condition. The overall care of a veterinary facility is essential to the delivery of excellent medical services.

SCENARIO

AD had just completed exercising the dogs in the outdoor runs. He came in through the kennel access door and pulled off his boots replacing them with the sneakers he had left on a bench near the exit. He went through to front office and the receptionist mentioned that the lab had just called to let them know they were faxing over the results of a patient's blood work.

AD noticed that there was no fax coming through, then checked the machine and realized that it was out of toner. Going to the supply closet, he picked out a new cartridge and put it into the machine so the fax would print. Then, he boxed up the old one to be recycled and added a fax toner cartridge to the front office want list.

Next, he went down the line of cages in the back, feeding each animal and clearing out waste. A small white dog sat in the back of one cage, shivering. When AD opened the cage door, the little dog raised his lip in a snarl. Leaning in, he patted him on the head and the little dog snapped at him, raking the back of AD's hand with his teeth. AD quickly closed the cage and went to clean the wound. Thankfully, the wound was not deep but the skin had been broken. To be safe, he checked with the veterinarian to see if there was anything else he should be concerned about.

- How did AD act proactively to create and maintain a clean, safe, and functional work environment?
- Which of AD's actions might have helped to spread pathogens among the animals in his care?
- What preventive measures could AD have taken to avoid the dog bite?

REVIEW

1. List the potential problems with leaving uneaten portions of canned cat food in a cat's cage.
2. List five significant steps that you can take to keep the facility clean.
3. What should you do if you notice that an exhaust fan in a kennel area is making an unusual noise?
4. List five potential safety hazards for employees in a veterinary practice.
5. List three precautions you can take to minimize the risk of injury by animals.
6. List five OSHA requirements for ensuring safety in the veterinary practice.
7. What can you learn from material safety data sheets (MSDS)?
8. List four types of information, other than MSDS, included in the safety notebook housed in a practice's right-to-know station.
9. List the protective items for use in radiology.
10. List five examples of capital equipment used by veterinary practices.
11. What purpose does rotating the inventory serve?
12. Name two categories of drugs that will not be found in the pharmacy cupboard.
13. List four zoonotic diseases, the source animal, and the method of transmission.
14. What are the steps you should take in processing orders when they arrive?

ONLINE RESOURCES

Veterinary Information Network (VIN)

VIN, an online service for veterinary professionals, offers a reference center, OSHA information, and classifieds for the veterinary industry. Membership is free for students at sponsored universities.

<http://www.vin.com>

Occupational Safety and Health Administration (OSHA)

Part of the U.S. Department of Labor, OSHA's mission is to prevent injury, illness, and death in the workplace. The Web site offers information on safety regulations, programs, services, and other resources related to workplace safety.

<http://www.osha.gov>

VetGuide.com

VetGuide is the illustrated buyer's guide to new equipment and supplies for the veterinary facility, offering links to a wide variety of veterinary products for both the doctor and the front veterinary office. This is good resource for learning about some of the capital equipment used in the hospital.

<http://www.vetguide.com>

Animal Care Equipment & Services (ACES)

This site features numerous types of animal care and handling equipment, and includes images of various cages, protective devices, safety vests, and veterinary accessories that an employee is likely to encounter in a veterinary office.

<http://www.animal-care.com>

CHAPTER 3

Front Office and Receptionist Duties

OBJECTIVES

When you complete this chapter, you should be able to:

- identify the types of records found in a veterinary practice, including laboratory reports, radiographs, medical records, vaccination certificates, and controlled substances
- describe the purpose of the various logs found within a veterinary practice, and who is responsible for maintaining the logs
- describe how to admit and discharge patients, take patient histories, maintain records, and prepare appropriate release forms and certificates for signature
- describe basic concepts of records retrieval and protection of medical and business records
- identify basic filing systems and filing equipment
- explain how to screen and process mail efficiently

KEY TERMS

reminder cards	master problem list	numerical filing
medical records	SOAP	alphabetic filing
signalment	purge	admitted
objective information	logbooks	consent form
subjective information	X-ray logbook	discharge
source-oriented	anesthesia record	neuter or spay
medical record	sheet	certificate
problem-oriented	controlled substance	vaccination certificate
medical record	logbook	health certificate
conventional format	laboratory logbook	

Introduction

Although the specific duties of a veterinary assistant, technician, or any member of staff assigned to the front office will vary slightly from practice to practice, you will undoubtedly be called upon to perform many clerical functions, such as maintaining and filing medical records, obtaining signed consent forms, admitting and discharging patients, sending **reminder cards**, and dealing with the mail.

Some of these responsibilities may seem a bit mundane on the surface, but they are vital functions that ensure the successful operation of a veterinary practice. **Medical records** summarize the patient's history for the veterinarian and provide specific facts about clinical signs of medical concerns, medications, and past treatments. Without these, the veterinarian would be working without a complete history which could contribute greatly to a diagnosis and treatment plan. Correct admission and discharge practices help to ensure that all concerns and issues are addressed before the patient is admitted or before the patient is discharged and the client leaves. Sorting and handling of the incoming mail ensures that important communications are received by the correct person as quickly as possible. Without efficient office management practices, a veterinary facility would be unable to function effectively.

Record Keeping

Purpose of Medical Records

Medical records are important documents that record the sequence of events each time an animal is seen by the veterinarian. They are also legal documents that contain pertinent information on each patient, as well as on each client, and should be treated with the same confidentiality as the medical records of a human being.

The primary purpose of veterinary medical records is to serve as a detailed description of medical issues, progress notes, treatment plans, options offered to the client, and the resolution of issues which may arise. Each record is a benchmark for measuring the improvement or decline of a patient's condition.

The medical record provides for continuity of care by reminding the veterinarian what she has seen on the physical evaluation of a patient, and of the patient's history, the tests that have been done and their results, the diagnosis, the treatment that has been initiated, and all comments and recommendations made to the client.

A medical record also provides continuity when multiple doctors are involved in a single patient's care. It allows any other veterinarian to quickly become informed of the patient's medical history, the current problems, and what treatment plans and options have been discussed with owner. Another way to think of the medical record, in this context, is as a summary of an animal's life from the perspective of its health, wellness, or chronic or acute disease and injuries.

Another important purpose of medical records is to facilitate rapid retrieval of information about patients. For example, the medical record allows a veterinarian to review the patient's major medical problems, its medical history, and behavioral history. A veterinarian can scan easily for episodes of illness, review the diagnostic and treatment plans, and learn the outcome of therapy. The consulting veterinarian has available, through an accurately kept medical record, all the wellness issues that have been addressed, including vaccinations, heartworm tests, surgery, lab work, and any concerns expressed by the owner on previous visits (Figure 3-1).

The second part of this form has been completed as an example of the type of information recorded for a typical wellness examination. Note that no area of the form has been left blank and that when findings are normal, it is recorded as such, leaving no question that an area might have been overlooked should a problem arise at a future date. The upper part is usually completed at the front desk when the client arrives. The veterinarian can then pull up the entire form on the computer terminal in the examination room and record all findings directly into the patient's record. As a courtesy, a copy may be printed for the client.

As a review, the top part of the patient history includes current owner information and the patient **signalment**, which provides identifying information for the patient. The second part of the form records the patient's examination and generates a history (documentation) of the initial visit and will be referred to in all future visits. As an example of the kind of information contained, it has been completed with the findings of what would be a typical wellness check or an incident of concern to the owner.

The medical record also provides rapid retrieval of information about a client and should include current, confirmed information regarding address and telephone number (including cell phone numbers), information about where the client works, and how he or she may be reached during the day. Quick access is especially important when an animal is in the hospital and a decision must be made quickly about a particular treatment or procedure.

A third purpose of medical records is to provide documentation of all medical decisions from a legal perspective. Medical records are legal documents and must be handled as such. You should record all pertinent information completely, accurately, and as soon as the information is available. In the event of any dispute, medical records provide clarification of what was done and why. They protect the veterinarian and staff by providing complete and accurate documentation.

Because of the legal nature of medical records, all information should be written legibly and in black ink—never in pencil. If a mistake is made, a single line should be drawn through the mistake and initialed (this indicates that an error was made rather than suggesting that information was changed). Then, the correct information should be entered immediately following the error. Do not erase, scratch out, or "whiteout" mistakes. Medical records should be maintained continually so that they may be reviewed at any time.

Yet another purpose of medical records is to provide the means to track certain data within the patient population. Veterinarians can collect statistical information about certain illnesses. For instance, how many cases of canine heartworm were detected in the practice within a certain time period or if there has been a sporadic outbreak of another communicable disease within a specific geographical area. A veterinarian or practice manager can also gather

FIGURE 3-1 A

APPLE CREEK VETERINARY HOSPITAL

Client ID:_____   PATIENT ID_____

Client Name_____   Patient Name_____

Client Address_____   Breed:_____

Telephone number_____   Species_____

Exam Date:_____   Weight _____

DVM: _____   Birthdate_____

                                                     Gender_____ S or N_____

FIGURE 3-1 B

Temperature: (*Normal) Temperature is 101.2*

Weight: *(Normal) body weight is within the range expected for this type of this breed 37.2 pounds*

Appearance: *(Normal) BAR*

Eyes: *(Normal) Eyes are bright and clear. The pupils are normally responsive to light*

Ears: *(Abnormal) Left ear has reddish discharge, odor. Yeast infection confirmed on microscopic examination*

Mouth and Teeth: *(Normal)Gums and teeth appear healthy. There is no significant buildup of tartar.*

Skin and Hair Coat: *(Normal) Healthy coat and skin. There are no signs of parasites, irritation or areas of alopecia*

Abdomen and Abdominal Organs: *(Normal) Abdominal organs/abdomen are palpably normal with no pain or tenderness noted*

Lymph Nodes: *(Normal) All lymph nodes appear normal on palpation*

Gastrointestinal: *(Normal) There is no reported history of vomiting or diarrhea. Appetite normal.*

Cardiovascular: *(Normal)strong, healthy heart sounds, valve sounds normal and without murmers Pulse rate/rhythm normal*

Respiratory: *(Normal) Respiration normal, lungs clear. Trachea appears normal*

Mucous Membranes: *(Normal) Healthy, pink CRT normal*

Musculoskeletal: *(Normal) Normal gait, no evidence of limping or joint pain, no heat or swelling. Patient is well muscled as typical of breed and excerise/activity level*

Urogenital: *(Normal): Palpation of bladder and kidneys is normal. There is no swelling, redness,or discharge*

Subjective, Objective, Assessment, Plan: *Owner concerned about reddish discharge and odor from left ear, head-shaking and scratching, sometimes crying when scratching. Laboratory results confirmed Malassezia (yeast) infection. Dispensed YXZ, 5 drops affected ear, BIDX10. Technician demonstrated how to perform ear cleaning, administer drops. Recommend patient is kept out of the water until infection has cleared and the use of "Swimmers' Solution" to help prevent further occurrence.*

FIGURE 3-1 Properly kept medical records will provide a complete history of the patient. This sample form is typical and may be similar to those found in the computer file. Regardless, all information about a patient must match exactly.

information about the profile of the overall patient base by utilizing medical records, determine the demographics of the practice and even the average number of pets per household.

Creating Medical Records

Each medical record must contain a certain minimum amount of information. For instance, the record should include the client's name, address, and telephone number, as well as the client's employer and work number, cell phone number, and the best way to reach the client during the day. Pertinent patient information includes the animal's name, species, breed, age, gender, reproductive status, color, and if the patient is a cat, it often includes hair type: long hair or short hair. Figure 3-2 shows a patient/client information form which would be completed on the first visit. It also includes patient signalment and clearly defines the authorization and terms of services provided.

A standard client form and authorization like this goes a long way to clearly define the terms of service and prevent any misunderstanding. This form is placed in the permanent client/patient file.

The most popular medical record format is a file folder with a folding clip at the top to hold 8½- by 11-inch pages in place. This design is very easy to use, whether making additional entries or retrieving information. There are different ways to identify the client and the patient on the outside of the folder. It is standard practice to print a client information label that adheres to the upper left hand corner on the outside of the folder and to put the first two letters of a client's last name on the right hand edge of the folder. These letters are bold, color coded, and make retrieval and re-filing easier. As with the client information labels, others are created for each of the owner's pets. These can all be created through the computer program in use at the practice.

Other notations can be included, but never write anything personal about the client or pet in or on the folder. Instead, the use of colored "dots" is an alert to anyone handling the patient or speaking with a client. For example, if a dog is known to be aggressive and a potential biter, a small red dot is often placed by the name label for the patient. This alerts everyone to be careful. Yellow might suggest caution and a black dot should always be added to a patient's label if the pet is deceased.

With each visit, all information should be confirmed that is still current and that the owner still has all the patients on file with the practice.

Much information, both subjective and objective, is contained in medical records. **Objective information** includes factual, measurable data, such as an animal's weight (Figure 3-3), body temperature, heart rate (HR), and respiration rate (RR).

Subjective information is an entry that describes the animal's overall attitude at the time of the examination. "Bright, alert, and responsive" (BAR) is a comment often included in this section of a medical record. Subjective information about a patient is not measurable, but observable. The subjective entries comprise the overall clinical impression of a patient, answering questions such as "How does the patient move?" and "How does the patient respond to its environment?"

WELCOME TO RIVERTON VETERINARY CLINIC

FILE # _____

Owner (s) _____Spouse _____

Address: _____

City/State & Zipcode _____

Main Phone # _____ Other phone/Cell# _____

Spouse's # _____Work Phone # _____

Driver's License# _____If recommended, by whom: _____

PET INFORMATION

Name: _____ Dog, Cat, Other _____ Breed _____

Color: _____ Different Marking(s), Tattoo, or Microchip #:_____

Approximate Age: _____ DOB:_____ Sex: Female/Spayed female or
Male/Neutered male _____

Where was your pet spayed or neutered? _____

Reason for visit: _____

AUTHORIZATION:
PAYMENT DUE AT TIME OF SERVICE
WE DO NOT BILL!

I understand that payment is required at the time services are rendered or upon release of the animal from the hospital. I agree to pay all reasonable attorney's fees and court costs, filing fees and commission of fifty percent that will be assessed to us by XYZ Recovery Services, retained to pursue in this manner. I further agree to pay interest at the rate of one and one half percent per month (eighteen percent per year).

Signature: _____ Date: _____

Spouse Signature: _____ Date: _____

© Delmar/Cengage Learning

FIGURE 3-2 A client/patient information form is an important part of the medical record.

The assessment of a patient, as well as formulation of the diagnostic and treatment plans, revolves around the careful compilation of subjective and objective data about the patient. These assessments are referred to as SOAP, an acronym for *subjective, objective, assessment plan.*

Organizing Medical Records

There are two primary ways medical records are organized: the **source-oriented medical record** and the **problem-oriented medical record**. Regardless of

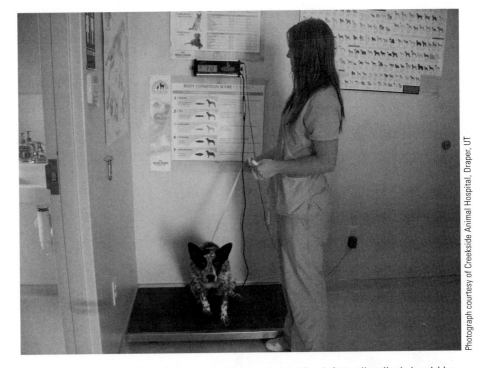

FIGURE 3-3   An animal's weight is an example of objective information that should be included on the patient's medical record.

the particular format, it is best to arrange information in the chart in reverse chronological order, with the most recent information on top, where it is most readily available.

The source-oriented medical record, also called the **conventional format**, can vary dramatically in size and detail. Data is entered as acquired, in chronological order. Often with this format, the records of multiple patients are kept in a single folder. Data for the different animals may be recorded on a single sheet, simply dated to indicate when each individual has been seen. This format requires much less time to complete than the problem-oriented format. However, compared to the problem-oriented system, it lacks the detailed documentation of procedures and their results, and it makes retrieving information more difficult. Maintaining adequate communication and legal protection is more challenging with source-oriented medical records. In contrast, the problem-oriented medical record format provides a complete, accurate, detailed chart on each and every patient. It offers a comprehensive review of the patient, including:

- the patient's history
- the patient's medical problems
- procedures or treatments that have been performed and the results
- a detailed plan for the patient's future

It is more time intensive to keep this style of medical chart. However, it improves communication among staff caring for the animals, as well as among doctors in a multidoctor practice. In addition, more of the patient's history and other data are readily available to support case planning and provide legal protection if a claim arises.

The problem-oriented medical record usually includes several standard sections. The first is the **master problem list** (Figure 3-4). Any issue requiring medical attention should be included on the problem list. Entries on the problem list might include an abnormality that the owner has noticed, an unusual laboratory finding, or a diagnosis. You can also think of the problem list as the medical history in brief or as an index to the rest of the medical chart. The problem list should also include the dates of inoculation, results of heartworm tests, drug allergies, ongoing medical conditions, and other information, depending on the standard practice of the office or hospital.

In addition to the master problem list, you may expect to find a place to record a comprehensive history, a form for recording the results of a physical examination, inserts for the results of laboratory analyses or special procedures, and progress notes (Figure 3-5). Progress notes are divided into four sections represented by the acronym **SOAP**:

- S—subjective
- O—objective
- A—assessment
- P—procedure or plan

Each problem is "SOAPed" separately. All communications with a client or any other veterinarian involved in an animal's care should be recorded in the progress notes.

The key to organizing medical records efficiently is to learn the format used in the practice where you work, become familiar with the details of the record-keeping system that is in place, and do all you can to help maintain the consistency of medical record keeping. This consistency ensures that each animal gets the best care by keeping all pertinent information about patients readily available for review. With the use of a computer system, the recording of all patient and client information has become much easier and more efficient. Patient notes are often entered directly into the computer into the patient's file in the examination room. The client information is entered directly into the computer file at reception. These computer files may be accessed from any computer within the network of the practice.

It is important to remember that both the hard-copy file and the computer files are exactly the same in the notes and comments entered. Ideally, we think of "going paperless" yet in practicality it is not always feasible to enter data directly into a computer file.

Maintaining, Retaining, Purging, and Releasing Medical Records

One of the most important aspects of medical record keeping is accuracy. It is critical that each medical record always reflects up-to-the-minute information

CITY ANIMAL HOSPITAL
Master Problem List

OWNER INFORMATION

Owner Name □ Mr. □ Miss □ Ms. □ Mrs.   Patient/Pet's Name

Address   City/State/Zip

Home Phone   Business Phone

PATIENT/PET INFORMATION

Chart # _____
Patient _____   Species _____   Breed _____
Color _____   Sex □ F □ M □ N   Birth Date _____
Vax History _____   Weight _____

IMMUNIZATION/PREVENTATIVE RECORD

DATE									
RABIES									
DA2PL									
FVR-CP									
FELV									
FECAL									

PROBLEM LIST

PROBLEM	DATE ENTERED	DATE RESOLVED
1.		
2.		
3.		
4.		
5.		
6.		
7.		
8.		
9.		
10.		
11.		
12.		
13.		

FIGURE 3-4 All issues and concerns need to be accurately recorded on the master problem list. This often travels with the patient file during the exam and treatment and updated into the computer file.

CITY ANIMAL HOSPITAL
Progress Notes

OWNER INFORMATION

Owner Name ☐ Mr. ☐ Miss ☐ Ms. ☐ Mrs. Patient/Pet's Name

Address City/State/Zip

Home Phone Business Phone

PATIENT/PET INFORMATION

Chart # _____

Patient _____ Species _____ Breed _____

Color _____ Sex ☐ F ☐ M ☐ N Birth Date _____

Vax History _____ Weight _____

CURRENT PROBLEMS

Date				Soap		
Month	Day	Year	Time	Format	Progress Notes	Fee
				S	Subjective Data	
				O	Objective Data	
				A	Assessment	
				P	Procedure for Diagnosis and Treatment	

© Delmar/Cengage Learning

FIGURE 3-5 Sample progress notes information sheet.

about a patient and a client. Here are some simple strategies for keeping medical records current:

- Ask each client who comes into the clinic if the address and telephone number recorded in the medical record are still correct.
- Ask the client about animals not seen in a while (another reason to always mark a patient as deceased to avoid what may seem to a client, an insensitive question).
- The veterinarian will record information in the medical record after seeing the patient and when completed, the files will be returned to the front office to be re-filed. The day an animal is presented to the practice, what is observed and what is done, as well as any assessments and plans, will be recorded in the medical chart.
- Front office and receptionists should also enter and record telephone conversations with clients in the medical record, even if a message had been left, noting the date and time.
- All sheets in the record should be organized with the most recent information on top.
- The master problem list to serves as an index or table of contents for the rest of the record.

Periodically, it is useful for the practice to **purge**, or relocate, the records of animals that are no longer active patients. Purging the inactive files reduces the amount of medical records in the file shelving by removing the charts of animals that have died, whose owners have moved, or that are no longer served by your practice—all creating better use of the filing system and available space. When a medical record has been removed from the active patient files, it should be marked as inactive and stored separately from the active files. Medical records should not be destroyed and must legally be retained for a period of seven years.

Every practice has its own policy on purging, but it is generally done at least once a year. Each practice also defines which records are considered active. For some practices, any animal that has been seen within three years is considered an active patient. Sometimes the time frame is as short as 18 months. The only way to do this is to look at each file, beginning at one end and by working through them "a letter at a time" until each file has been checked, a task that can be easily divided between staff members to work on during downtime.

If a patient has died or the owner has moved away, the file can be removed from the active records. The hard copy of the medical chart should then be stored by the practice in case there is ever any reason to retrieve information about the animal or the case. If the pet simply has not been seen for a long time the medical record is usually considered inactive. Inactive files are best stored where they can easily be retrieved if the client wishes to bring the same pet or perhaps a new one back to the practice for medical care.

Prior to inactivating a medical record, most practices contact a client by phone or in writing about an animal that has not been seen for a certain period of time. A contact letter might read as shown in Figure 3-6.

A letter like this gives a client a chance to respond. The animal may have died or the owner has moved away. Often these contact letters are

---

CITY PARK ANIMAL CLINIC
2765 Park Center Dr.
City Park, WA

Dear _____,

We are currently updating our files and noticed that we
have not seen _____(pet's name) since _____
(date). We would appreciate just a few moments of your time to
help us keep your pet's medical records current. Please contact
us with any concerns you may have or how we may be of
assistance to you. 1-800-555-4344

Thank you, (signature),

Receptionist

FIGURE 3-6   A contact letter may be sent to a client before inactivating
a patient's record.

returned as undeliverable which indicates that the client no longer lives at
the address on file.

Another method of client contact is to send **reminder cards**, preprinted
post cards that remind owners when yearly vaccinations are due. These can usually be generated by the computer system and sorted by the month prior to when
the vaccination are due. A simpler and more cost-effective method would be to
contact clients via e-mail. If your practice does not ask for an e-mail address, it
would be a good thing to consider adding to the client information sheet.

A veterinary medical record is a confidential document. The record is
owned by the veterinary practice, not by the owner of the animal. Most
practices require a signed authorization form from the owner (or an authorized agent) before any information from the medical record is released to
another veterinarian, a third party, or even to an insurance company. The
attending veterinarian is usually the only person authorized to release information from the medical record. Figure 3-7 is an example of a medical
disclosure form.

Occasionally a request will come from a public agency, such as a local animal control office or the police for the vaccination status of an animal that has
been involved in a bite injury to a human. This is an important exception to
the general rule barring the release of confidential veterinary medical information. If you have any questions about releasing medical records in a particular
situation, consult the veterinarian.

When the medical record is being transferred to another veterinarian, it
should be sent directly to him or her. The original medical record is never

---

CONSENT TO DISCLOSURE OF MEDICAL RECORDS

I, (name of owner or authorized agent), the (owner or agent) of (name and breed of animal), understand that the information contain in (name of animal's) medical records is confidential. However, I specifically give my consent for (name of practice) to release all medical records concerning (name of animal) to (name of the party to whom information is being released).

The medical records are to be disclosed for the specific purpose of _____.

It is further understood that the information disclosed may not be given, in whole or in part, to any other person than stated above. _____

_____

Signature of owner or authorized agent

_____

Date

© Delmar/Cengage Learning

FIGURE 3-7 It is essential to complete a medical disclosure form. It is a waiver of confidentiality and be completed before the release of any information.

released—only copies. When the transfer of a medical record is requested, that request should be recorded in the original medical chart. Clients may also request copies of their pets' medical records, especially when they are moving out of the area. Part of the patient's history may include radiographs and with these too, only copies of the films should be provided. Generally, there is no charge to the client for paper copies of medical records; however, because of the expense involved in making a professional copy of an X-ray film, many practices charge a nominal fee to the client.

Business Records and Logbooks

The business records for a veterinary practice are usually kept separate from the medical records, although there is a bit of overlap in some cases. For instance, veterinary-specific software now links invoicing to the medical record and history. Most, if not all, practices keep business accounts with a computer program that is separate from their management of medical information and client invoicing.

Computerized business records save time and provide an enormous volume of useful information. They empower a practice to identify trends within the practice, track client purchasing choices, control inventory, and evaluate income versus expenses. Although you may be asked to learn to generate invoices for individual transactions with clients, most business records are attended to by the practice manager.

Unless there are extraordinary circumstances, almost all practices now operate on the simple premise of "Payment Required at the Time of Service." Payment may be made in cash or by a credit card or a written check that is accepted with the use of a check verification system. This "no credit policy" not only saves time by not having to send monthly statements to clients but saves money as there is no postage. This policy dramatically reduces financial loss to the practice. There are also credit cards available which are specifically limited to use for veterinary fees. Your practice may or may not participate in one of these programs, but if so, the application process is conducted through the practice manager or head receptionist. This application is confidential and the telephone application/approval process must be conducted away from the front office and in private. In most circumstances, this specific credit application will either be granted or denied within minutes. Do not discuss the application when other clients are present. It is best to use an available exam room for privacy. Never offer credit to a client, regardless of how well known the person may be to you personally or to the practice. Extending credit or account privileges is solely at the discretion of the veterinarian or practice owner.

Part of the business record-keeping system in any veterinary practice consists of various **logbooks**, where medicine and business overlap. Logbooks of any type are maintained as hard copies. One example of a logbook used in veterinary medicine is the **X-ray logbook** (Figure 3-8), which includes the following information:

- client's name
- patient's name
- date
- patient ID number
- body part to be radiographed
- view(s) taken
- measured thickness of the body part(s) radiographed
- X-ray machine settings used for each view
- comments regarding the quality of the films: too light, too dark, or background contamination, in short, any abnormality of the film quality should be noted.

The X-ray logbook is maintained by the technician taking the radiographs and is normally kept in the radiology room. The diagnosis or significant findings are evaluated and recorded by the attending veterinarian who will enter the information into the patient's medical records.

Another example of record keeping is the **anesthesia record sheet**. This documents the date, the patient and client, the procedure(s) performed, all drugs administered including the time, the exact volume of drugs given and the routes of administration, heart and respiration rates, blood levels of oxygen saturation, the duration of the anesthesia, including time of induction and the length of the procedure, and the names of the surgeon(s) and anesthetist. The anesthetist should also enter notes at the time any unusual response or event occurs.

X-RAY LOGBOOK

Date	Case Number	Patient's Name	Client's Name	Body Part	View(s) Taken	Thickness of Body Part	X-ray Machine Setting

Comments:

FIGURE 3-8 A sample of an X-ray logbook and the information required. This is normally completed by the veterinary technician who takes the radiograph but it is kept with the patient records.

The anesthesia record sheet is maintained by the anesthetist and recorded by him or her during and throughout the period of time the patient is under anesthesia. This very important document details minute by minute, along with continual monitoring by the anesthetist, the entire sequence of the anesthesia.

Every veterinary practice is also required by law to have a **controlled substance logbook**. The controlled substance logbook is important from the perspective of preventing and/ or alerting the veterinarian to the potential of human drug abuse either by clients requesting refills frequently or, unfortunately, potential drug abuse by staff or anyone having access to the controlled drug cabinet. The law requires accountability for the amount of drugs ordered and the amounts dispensed. Because some of the drugs used in veterinary medicine have the potential to be abused, they are regulated by the Drug Enforcement Agency (DEA), a branch of the federal government. The DEA has the authority to perform random inspections, including examination of the controlled substance log, the matching up of inventory on hand against recorded dispensed amounts, the method in which these drugs are secured, and who has access to the controlled drug supply.

All details of controlled substances must be recorded in the controlled substance logbook. First, the delivery of an unopened order to the central supply of the practice's pharmacy is recorded in one part of the log. Then, when a container is opened to dispense medication to a patient, that action is recorded on a different page of the log. Every tablet, capsule of every controlled substance must be accounted for in the logbook. Every milliliter of a controlled injectable medication must also be recorded.

The controlled substance logbook should include the following five items:

1. drug dispensed
2. date, client's name, and patient's name
3. amount dispensed and the amount remaining
4. name of veterinarian authorizing the use of the medication
5. the individual actually handling the medication

Inventory is required every two years, and the amount remaining in the logbook must agree with the amount in the controlled substance cabinet. Medical record entries should also agree exactly with the information in the controlled substance log. Controlled substances are to remain locked up at all times in a secure cabinet that should be mounted permanently in the practice and separately from the general pharmacy cupboard. Usually, the only staff members of the practice who should have a key to the controlled substance cabinet are the veterinarians involved in the practice as ultimately, they are responsible under the law.

Most veterinary facilities have a rule: "If you use it, you log it." This rule simply means that if you are the one instructed by a veterinarian to retrieve a controlled substance for dispensing purposes or for use on a hospitalized patient, you are the one who is to record that event in the controlled substance log. You must learn and become comfortable with the process of recording the use of these medications. *Falsifying the controlled substance log or improperly using any controlled substance is a federal offense.*

The final example of a logbook used in veterinary practice is the **laboratory logbook**, which provides a synopsis of the laboratory tests that have been ordered. At the very least, the date, client and patient names, the technician performing the tests and whether the tests are run in-house or sent to an outside laboratory, the date and the lab to which the samples were sent are all recorded in the laboratory logbook.

This logbook is kept in the laboratory area and the person handling the samples or performing the tests is responsible for accurate recording of the information. While you may not be performing the test yourself, any sample you take to the lab should be recorded and initialed by you. The technician can check through the logbook and see which tests are pending.

Filing

Filing in a veterinary office is much like filing in any other office setting. The details vary slightly from practice to practice, but the concepts are universal. Related materials (medical records, for example) are filed in the same location. For easy retrieval and re-filing, they are usually in the front office in the vicinity of the receptionist's desk. They must not be housed in an area that is available to the public. The invoices for products and materials purchased by the practice, as well as other expenditures, are filed in a different location than the medical records—most likely in the practice's business office. Client information and handouts are usually filed in a different location, perhaps in a file drawer in the front desk or in lateral file cabinets (refer to file systems, below).

Filing Equipment and Supplies

Most practices have a variety of filing equipment available, depending on the items that need to be filed. For instance, if medical records are kept in manila end-tab folders (designed to hold 8½- by 11-inch pages), there will be open shelving to accommodate the files. If the medical records are kept in hanging pocket folders, there will be cabinets with pull-out drawers.

In order to maximize the usable space within a reception area, file cabinets with lateral file drawers are often used (Figure 3-9). When the drawer of a lateral file is pulled out, it does not take up as much room as a standard file cabinet drawer. The files are then arranged from side to side rather than from front to back.

The filing supplies found in a veterinary office will vary as well, depending on the filing needs of the particular practice. Most medical records are identified on the outside of the folder or pocket using letters, numbers, or a combination of the two. There should be a supply of blank folders (usually with end tabs) or hanging pockets, rolls of number stickers, and rolls of letter stickers. In addition, there should be a supply of side-tab manila folders for more conventional filing. The manila folders may rest directly in a file drawer, or they may be inserted into hanging files. There will be stickers for identifying the manila folders.

Filing Systems

You may encounter different filing systems, depending on the practice where you work and what kind of information retrieval systems the practice has developed. Medical records are usually filed in one of two ways: numerically

Photograph by Kathy Nuttall, courtesy of Riverton Veterinary Hospital, Riverton, UT

FIGURE 3-9   An example of neatly organized client/patient files using a lateral file cabinet.

or alphabetically. A **numerical filing** system can work in one of two ways: either each client is assigned a number, or each individual patient is assigned a number. Each digit, from 0 through 9, is represented by a different color. The colors on the records then vary, depending on the number that has been assigned to a particular client or patient.

As an example, a practice using numerical filing of medical records might use end-tab folders and apply the colored numbers to the end (or bottom) of the folder. Usually a single sticker with the last two digits of the year is applied on the bottom edge of the same tab. Then the medical records are filed on open shelves with the end tabs, complete with numbers, visible to the receptionist. It is easy to see at a glance if there are files out of sequence. Tabbed dividers are very helpful in clustering together records with similar number sequences. This system is time consuming as first the client name has to be located in order to retrieve the correct number.

For medical records filed using the **alphabetic filing** system, the concept is very similar to numerically sequenced medical records. Each letter of the alphabet is assigned a color, and the first two or three letters of the last name are attached to the right side tab of the folder: for example JONES would be designated JO while Mc Whitten would receive three letters, McW.

The practice where you work will probably have filing systems in place for more than just the medical records. For instance, there are probably client education handouts that are frequently used in the practice. There will also be various forms and certificates, which are used throughout the practice that must be kept organized and easy to find. One strategy is to have hanging files organized by topic, and then, within the hanging files, to have manila-tabbed folders organized alphabetically. This system allows related information or paperwork to be stored in the same place and to be organized and accessed easily.

All the forms that are used and reproduced within the clinic should have one file containing all the master sheets for making additional copies. If a yellow highlighter pen is used to write "master" across the retained copy, it will prevent the last copy from being used, acts as a reminder to print more copies and the yellow highlighter will not be visible on the new copies.

Making Filing Efficient

When you are getting ready to replace files in their appropriate places, use your time as efficiently as possible. The first step is to separate the files into their appropriate groupings. For instance, gather the medical records together, the client education handouts together, and so forth. Next, be sure that all the pertinent and updated information is in each of the files. You should check to be sure that all appropriate entries have been made in the medical records before they are returned to their shelves.

When you are confident that updating is complete, you are ready to begin the filing process. If you are filing medical records folders, arrange them in the appropriate order, either numerically or alphabetically, before filing them. One time-saving trick when retrieving a file folder from open shelving is to pull out the adjacent file a short distance. This simple technique marks the location from which you have retrieved the file and makes it easier to return. It is best to re-file medical records frequently throughout the day to eliminate clutter and to make the records easy to locate if they are needed later.

Admissions and Discharges

Admissions to and discharges from the veterinary practice are usually completed by a veterinary technician or the receptionist. Front office staff are there to greet the clients and to admit the patient, either for a drop-off examination or surgery that day. Discharges are similar but usually follow a consultation with the veterinarian. You should know the procedures for admitting and discharging patients as this is an integral and important part of your job. Most practices have clearly outlined procedures to follow—usually a step-by-step process that includes specific paperwork (Figure 3-10). It is important to document all procedures that are to be performed on an animal, as well as any medication the animal is to receive upon admission for a procedure.

An important part of the admission process is having written consent to proceed with diagnostic procedures, surgery, or with a treatment plan. *Upon discharge, it is equally important to have written discharge orders with clear follow-up instructions when the patient leaves the hospital.* This allows the client to review what she has been told at the time a patient is discharged and helps to prevent any confusion from inaccurate recollection of verbal instructions.

© Delmar/Cengage Learning

FIGURE 3-10 Proper documentation is important for both admissions and discharges.

Admitting a Patient

Animal patients are **admitted** to a veterinary hospital for many reasons:

- The client may be unable to stay for a procedure that would normally be performed on an outpatient basis.
- The pet may be scheduled for a wellness examination, or perhaps it is sick or injured. It is especially important for you to get detailed information from the client when pets are "dropped off." You must know exactly why the animal is being admitted, the procedures scheduled, and especially how to contact the client during the day.
- The animal may be admitted for a period of supervised care while the client is out of town.
- The patient is being admitted for a surgical or dental procedure.
- Perhaps the animal is boarding at the veterinary hospital and having a medical or surgical procedure done during its stay.
- Certain laboratory analyses may take more time than those performed on an outpatient basis while the client waits.

Regardless of the reason, it is important to know precisely why the animal is being admitted and have all required paperwork completed before the client leaves.

It is critical that you get all the relevant information from the client at the time of the admission. You must be sure to take a thorough and accurate history. Record all concerns or questions the client has and confirm how and when the client may be reached. Record telephone or pager numbers in case the client must be reached to answer questions, to give consent for additional procedures, or simply to be updated about the patient's condition.

An animal should never be left at a veterinary practice without a written, signed **consent form** in the medical record. By reading and signing the form, the client gives his or her informed consent for the procedures to be done and acknowledges the potential risks. Most hospitals use either preprinted fill-in-the-blank forms, or they have prepared consent forms, loaded into the computer system, that may be personalized with specific client and patient information. Most forms tend to have fairly standard consent language. You will need to become very familiar with the forms used by the veterinary practice where you work, because you may be asked to answer clients' questions about giving their consent.

You should be able to explain, step by step, each section of a consent form and without going into surgical details have some basic knowledge of the procedure listed. For example, a surgical consent form may include possibilities that may occur during the surgery, such as intravenous fluid therapy or an option for postoperative pain management. You need to know what your veterinarian recommends and why. When the consent form is signed, the client gets a copy for his or her own records, and the original goes into the medical record. A surgical/anesthesia consent may also be a separate, signed document which is as straightforward and simple as the following (Figure 3-11).

For a nonsurgical admission (e.g., a patient who needs an internal medicine workup), the veterinarian will list on the consent form all procedures she anticipates doing. Sometimes, the form uses general language like "appropriate laboratory tests." Once again, the client gets a copy of the signed consent form, and the original is inserted into the medical record.

Discharging a Patient

Often, the veterinarian will meet with the owner before the **discharge** of a patient. This may be after a period of hospitalization in order to review with the client test results, surgical findings, postoperative care, or recommended follow-up instructions (Figure 3-12). After the consultation, the client and patient complete the discharge procedure at the front desk.

SEDATIVE/ANESTHESIA CONSENT

You are to use all reasonable precaution against injury, escape, or death of y pet. I understand that all anesthesia involves some minimal risk to my pet, but you will not be held liable or responsible in any manner whatever or under any circumstances in connection therewith as it is thoroughly understood that I assume all risks.

I have read the foregoing and agree.

_____   _____

Owner or Responsible Party   Date

FIGURE 3-11   An example of a consent form.

CITY HOSPITAL
ANY STREET
ANYTOWN, SS 00000

FOLLOW-UP INSTRUCTIONS FOR HOME CARE AFTER SURGERY

Due to the effects of general anesthesia, your pet may appear more tired than normal and possibly a little uncoordinated. This is to be expected, and the grogginess should disappear in a day or two, at the latest.

- To prevent vomiting due to excitement upon arriving home, do not give your pet food or water for an hour after returning home. Feed the regular diet lightly today, then return to normal feeding tomorrow.

- If your pet becomes listless or refuses to eat the next day, or if vomiting or diarrhea occurs, call the veterinary hospital immediately.

- Discourage your pet from licking or chewing the stitches or incision. Call us if licking or chewing persists.

- Check the incision site twice a day for any redness, swelling, pain, or drainage. Report any of these immediately.

- Limit exercise for a week following surgery. (No running or jumping)

- Keep any bandage or cast clean and dry. Contact the veterinary hospital if the bandage is excessively wet or dirty.

FUTURE TREATMENT

☐ Please make an appointment for a recheck in _____ days/weeks.

☐ Please make an appointment for suture removal in _____ days.

☐ Please make an appointment for bandage change or removal in _____ days.

☐ Remove bandage at home in _____ days.

ADDITIONAL INSTRUCTIONS:

_____

_____

_____

_____

_____

_____

© Delmar/Cengage Learning

FIGURE 3-12 It is important to provide written follow-up instructions for patients being discharged.

The practice may use standard written discharge instructions to follow what are considered routine procedures such as a spay or neuter. These instructions may be stored in the computer to be personalized for each patient, or they may be preprinted forms. Most discharge orders have blanks which are filled in and most include comments about the following:

- prescribed medications, including the drug name, the dosage, and how often it is to be given
- dietary modifications
- the patient's recommended activity level
- bathing or being in water, playing with other pets, or the level of necessary confinement
- care of an incision and when the sutures should be removed
- specific instructions about what to look for and when to call if a problem arises
- scheduled follow-up visits

When the client has reviewed the discharge instructions and has had his or her questions answered, a copy of the signed discharge form goes with the client and the original remains in the medical record. At the time of discharge, the client often receives a copy of any laboratory analyses that have been performed.

Forms and Certificates

There are many different forms and certificates in use at a veterinary practice. Become familiar with all of these in case you are asked to fill one out or to answer a client's questions.

We discussed, in the previous section, consent forms and discharge instructions. Other forms may be created for the client at the time a patient is discharged from the hospital. A **neuter or spay certificate** allows the client to prove that the animal is no longer sexually intact. This proof may allow the client to get a discount on the pet's license and may be required for the client to fulfill his or her obligation to a shelter from which the animal has been adopted.

Vaccination certificates are usually prepared each time an animal has received an inoculation. Rabies certificates are handwritten on county issued forms, printed in triplicate, and numerically sequenced. With each certificate, a numerically matched metal tag is issued to the owner and most counties require the tag to be clearly visible and attached to the pet's collar. A licensed veterinarian is the only one authorized to sign the rabies certificate. Other vaccination certificates may be generated from practice's computer program and linked to invoices so that they are automatically printed any time the animal receives a vaccination. Computer linking can also generate future reminder cards for inoculations, as well as help identify a lost animal by matching the number of the rabies tag to the owner's name and address.

Health certificates are more complex forms that are required for interstate and international transport of an animal. The veterinarian is the only person authorized to sign a health certificate.

Most of the certificates used in veterinary practice require a veterinarian's signature. If the form is not required to be filled out by only a veterinarian, then you may be asked to prepare the certificate ahead of time or while the patient is in the hospital for the appointment. In these cases, you will give the form to the veterinarian for his or her signature when the information contained on it is as complete as possible. If the patient has been hospitalized, you may be asked to place the certificate on the doctor's desk or in an internal mailbox or message box, or you may simply be asked to have the form ready for signature at the front desk when the veterinarian escorts the client to the reception area.

Every practice has a slightly different routine to be followed for ensuring that record keeping, certificates, and other appropriate paperwork are handled in a timely, efficient, and accurate manner. It is important to learn the details of how things are done at the practice where you work.

Screening and Processing Mail

It will no doubt be a part of your front office tasks to handle the practice's mail, both incoming and outgoing. You need to know how the mail is sorted and delivered to the appropriate person. If unsure, always ask for the specifics as they apply to your particular practice.

Incoming Mail

You will need to know the preferred process for screening the incoming mail in your practice. Be prepared to be able to answer the following questions:

- Is the mail delivered to the practice facility by a mail carrier, or must it be picked up at the local post office?
- Is collecting the incoming mail the specific responsibility of one person in the practice, or is it a shared job?
- What time of day is the mail delivered, or by what time can you count on the mail to be ready in the post office box?
- What time can you expect alternative deliveries like UPS or Federal Express?
- Who may sign for deliveries of mail or packages that require a signature?

Undoubtedly, there will be a plethora of supplier catalogues, professional journals, and magazine subscriptions for clients and children to enjoy while in the waiting room. All professional journals should be delivered to the veterinarian or placed in the staff lounge, never left in the reception area. The same applies to supplier catalogues. With the general subscription magazines, these should be placed in the reception area, removing the older copies and replacing them with the new issues.

Next, sort the mail by addressee (recipient), and then place each piece into the appropriate receptacle. The practice may have trays marked with each staff member's name for depositing phone messages and mail. Who receives envelopes addressed to the practice (the practice owner, the clinic manager, or someone else)?

You must also learn the process for accepting deliveries of medical or office supplies to the practice. Where do the boxes go when they arrive at the practice?

Who is responsible for unloading shipments of medical supplies? Is there any way you can assist that person, either by reviewing the packing slip to be sure that everything has arrived or by helping to put things away and rotating the stock? How are packages to be handled that read "Refrigerate Upon Arrival"?

Every practice has its own ways of dealing with incoming mail and packages and once you have learned the established routine you will become very proficient and helpful to everyone. You should never open mail that is addressed to a specific person no matter how inconsequential it may appear.

Outgoing Mail

When dealing with outgoing mail, the first step is to classify it for postage and shipping preferences. First-class letters receive different postage than reminder postcards directed to clients. Packages are often sent though delivery companies (UPS, Federal Express, etc.) and there may be specific envelopes or boxes and a telephone numbers to call for pickup. You will need answers to the following questions:

- How is postage for the outgoing mail handled?
- Is there a postage scale in the practice for weighing mail and determining the appropriate postage?
- Is there a postage meter in the practice for applying postage without using stamps and without having to take the mail to the local post office?
- Is the outgoing mail picked up at the practice by the postal carrier or does it need to be taken to the local post office?
- Which person in the practice contacts UPS, Federal Express, or other alternative carriers for a pickup?
- Will you be authorized to prepare outgoing packages for pickup?

Handling Mail Problems

Different types of problems with letters, packages, and carrier services sometimes occur. For instance, if a letter or a reminder postcard to a client is returned because the address is no longer correct, you need to know how to track down the new address and amend the information in the client's file, both the hard copy and in the computer. Sometimes the forwarding order has expired and the postal service is no longer delivering mail to the new address after it has been sent to the previous address. Very often, however, there is a sticker on the front of the card or envelope that lists the new address. It is easy to update the computer record and client file at that time. Sometimes a client receives mail at a post office box rather than at a physical address. This problem is relatively easy to avoid by carefully filling out the client information form or computer screen when the client first comes into the practice. Most new clients distinguish between the physical address and the mailing address.

If clients ask the practice to send a product or medication to them because it is inconvenient to stop in to pick it up, you need to know how to handle the details. Whose responsibility is it to decide what can be sent through the mail and what cannot? How are medicines packaged? How are the clients charged?

Most problems with postal or package services can be resolved easily and by knowing which member of staff is most likely to have the answers to the problem. You need to know who has the authority to make decisions about incoming and outgoing mail and packages. You will then be in a position to know how you can be most helpful in processing the large volumes of mail that pass through the veterinary office.

SUMMARY

Medical records are the single most important tool a veterinarian has, providing him or her with the patient's medical history, wellness examinations, and vaccination history. They provide documentation relating to clients, their animals, tests and procedures, and medications, and they are essential from the standpoints of health, efficiency, and legality. Veterinarians rely on their office management team to ensure that medical records and communications are properly labeled, filed, and organized in an efficient, easy-to-understand manner.

The office procedures within the veterinary practice is, in many ways, just as important as the medical procedures. As a staff member, you will likely be responsible, at least in part, for maintaining client invoices, medical records and logbooks, forms and certificates, admitting and discharging patients, handling incoming and outgoing mail. While these duties may seem tedious at times, even the slightest error can have far-reaching consequences, so it is essential that you are knowledgeable and skilled in performing your duties.

SCENARIO

A client of several years has brought her 6-year-old, long-hair cat in for an appointment after realizing that the cat was vomiting almost daily. She said that the "vomit was the same color as her cat food—only hairy." She was greeted by the front office staff and as the client's file had already been pulled, she was taken into the examination room to update the file on the computer terminal in the room and have all the information ready to meet with the veterinarian.

A veterinary assistant held and petted the cat while the veterinarian reviewed the file and asked the client some questions about the cat's recent health and behavior. The veterinarian who had not seen this patient before read in the file that this cat had been brought in one month ago for a similar problem.

By reading the history and asking several questions of the cat's owner, the veterinarian was able to determine that this was a repeat episode of hairballs, due to the cat's normal grooming behavior and ingestion of loose hair. He explained the situation to the client who then remembered the other episode and agreed that this was the same sort of thing and then apologized because she hadn't been brushing the cat or giving it the recommended treatment as often she probably should have. The problem was easily addressed and the client reassured, without blame, that this is a fairly common problem with long-haired cats.

- How did the use of medical records play into this scenario?
- What might have happened if the medical record had not been available?
- How did proper filing aid the front office staff in this situation?

REVIEW

Indicate whether statements 1–7 are true or false. If false, what would be required to make them true?

1. Medical records provide for rapid retrieval of client information as well as patient information in case of an emergency.
2. Medical records allow a veterinarian to track certain medical trends within the patient population.
3. All information in a medical record should be written in pencil in case important changes need to be made.
4. After the client signs the consent form, the original should be given to the client, and a copy should stay in the medical record.
5. Upon admission of his or her animal, a client can give verbal consent for the staff to proceed with diagnostic procedures and treatments.
6. A surgical consent form should include an explanation of the risks of anesthesia.
7. When sorting mail, you should open all envelopes with a letter opener, including personal or confidential mail.
8. List 10 components of a medical record.
9. Who is authorized to release information from a medical record?
10. How long should inactive medical records be stored?
11. List three precautions that you must take when recording information in the controlled substance logbook.
12. What are the two systems commonly in use for filing medical records?
13. What different types of medical records are mentioned in the text?
14. List three ways to make filing more efficient.
15. List five considerations for a patient that a discharge form might address.
16. What are two important types of certificates that require the signature of a veterinarian, and describe their purpose.
17. List three questions you should ask to learn the procedure for screening incoming mail.
18. List two alternatives to the U.S. Postal Service for picking up outgoing mail.

ONLINE RESOURCE

This Web site offers a variety of preprinted forms, labels, and filing system products designed specifically for the veterinary office.

http://www.ancom-filing.com

CHAPTER 4

Computers

OBJECTIVES

When you complete this chapter, you should be able to:
- identify the basic components of a computer and explain the function of each
- explain what software is and give examples of the type of computer program most often used in a veterinary office
- describe different types of computer networks and how they are used
- describe common features of veterinary practice Web sites and how they can benefit and improve aspects of practice management

KEY TERMS

software	toner
central processing unit	operating system
hard drive	word processing
CD-ROM	network
CD-R	intranet
monitor	workstation
mouse	server
keyboard	extranet
printer	firewall
modem	Internet
scanner	Internet service provider
USB flash drive	database
thumb drive	broadband
digital camera station	Web site
hardware	

Introduction

Veterinary offices, like other medical practices and hospitals today, rely on computers for scheduling, organizing data, tracking patient information, accounting and billing, messaging, and the daily organization of the practice. Although you probably have some experience working with computers, you may not have had exposure to specifically designed veterinary programs. Different practices may vary in the computer program used but all of the veterinary **software** programs, those that direct the operation of the computer, have been created for ease of use and may be easily customized to meet the needs of a particular practice. If you have never used a computer before, it may seem a little intimidating at first. Computers and computer programs are not designed to be difficult but rather to make life easier. Once you have developed a basic understanding of computers, you will find it much easier to use one efficiently. New employees are given extensive training in how to use the office computer and given every opportunity to practice and learn how to navigate and use the system confidently and correctly.

There is a wide variety of computers and computer programs available yet they all have basic similarities. On the job, you will learn to use your employer's particular software programs and you will be expected to understand some general facts about computers and their components.

Computer Components

The setup of all computer systems is basically the same. The computer system in a veterinary practice usually consists of a **central processing unit** (CPU) with a **hard drive, CD-ROM** or **CD-R** drive, a **monitor, mouse, keyboard**, and at least one **printer**. It may also include a **modem** (if your office has Internet access), a **scanner**, external storage devices such as a **USB flash drive ("thumb drive")**, and a **digital camera station**. The physical components of a computer system are known collectively as **hardware**. Figure 4-1 shows the components of a computer system.

The printer is an especially important piece of equipment, as you will be using it to print reports, client information sheets, reminder cards, statements, copies of patient medical records, invoices, and receipts. You will need to know how to load the paper and clear a "paper jam," change the **toner** cartridge and drum, or printer ribbon, and become proficient with printer maintenance and to troubleshoot problems which may arise. In addition, many printers now have the capability to scan and fax information (Figure 4-2). Experienced staff will provide you with the training and practice you need to perform all these functions.

Generally you would not need to know a great deal about the hardware beyond turning everything on and off correctly. Most of the training you receive will be for the software programs used with the computer system. Software may include the **operating system**, **word processing** programs, Internet and e-mail applications, and the specific veterinary programs used in the practice.

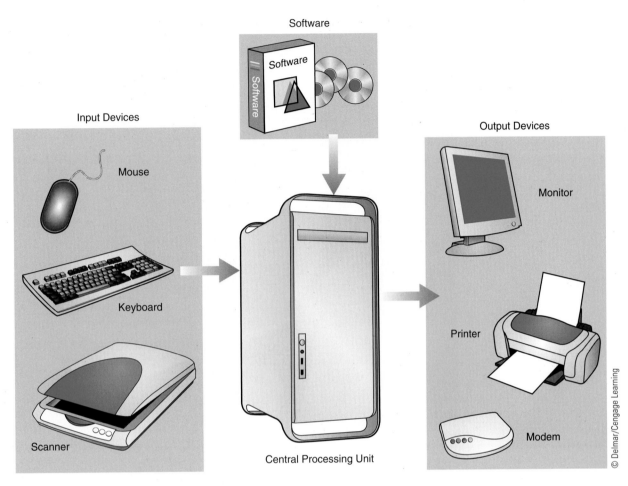

FIGURE 4-1   Various components of a system.

© Delmar/Cengage Learning

Veterinary Software

Computer software can simplify most of the tasks in a veterinary office or animal hospital. There are a number of veterinary computer programs on the market today, used for a variety of purposes, including:

- organizing medical records
- scheduling
- inventory management
- marketing
- generating performance evaluation reports
- time card processing
- posting account ledgers for inventory and clients
- merging new files into an alphabetical or numerical filing system
- creating invoices
- formatting, storing, and printing form letters, mailing labels, payroll summaries, paychecks, charts, graphs, bank statements, telephone directories, and other data

Photograph by Kathy Nuttall, courtesy of Riverton Veterinary Hospital, *Riverton, UT*

FIGURE 4-2   Modern office equipment often combines printer, scan, and fax capabilities into one compact machine.

The ease of performing these tasks on a computer explains why most receptionists and veterinary assistants feel fortunate to work in a computerized office.

Commonly used Programs

There is no single industry standard program that you will find on every computer, but here are some common types of programs and examples of each:

- operating system: Microsoft Windows, Mac OS
- general word processing: Microsoft Word, WordPerfect
- general database management: Microsoft Access, Corel Paradox
- general spreadsheet: Microsoft Excel, Corel Quattro
- general communication (including e-mail): Microsoft Outlook, Windows Live Mail
- Web browser: Internet Explorer, Mozilla Firefox, Google Chrome

Your office will probably use a combination of several of the programs and may update sites and preferences occasionally. All staff will receive training, often from a representative of the new program, explaining fully how to use the system, its features, and benefits.

Computer Networks

Unless you work in a very small practice, there will likely be more than one computer in the office. Veterinary hospitals and practices with multiple computer stations often employ a computer **network** system that links the computers together. There are three basic types of computer networks:

1. *Intranet.* An intranet is a computer network that links only the computers in a single office or building. Only authorized computer users in the building can access the intranet, using an assigned user name and password. Each computer on the network is called a **workstation**. Each workstation is wired into a central hub, which allows the workstations to "talk" to one another. When there are more than a few computers in the office, there may be a central **server**, a computer dedicated to managing network traffic and resources. The server might be used to store and manage common files, process print jobs for shared printers, and maintain a shared Internet connection.

2. *Extranet.* An extranet serves essentially the same function as an intranet, but allows limited access to people off-site. A user name and password is required to gain access to the extranet which is normally protected by a **firewall**, software or hardware designed to limit access to the server. Extranets are most often used by large companies with employees who frequently work off-site or by organizations with multiple locations and need to share information. Extranet use is usually accessed via the Internet.

3. *Internet.* The Internet is an enormous worldwide series of computer networks that allows individual computer users to access innumerable personal or business servers. Although a user name and password is required to gain access to the Internet through your **Internet service provider** (ISP), you generally do not need them to gain access to most sites (Web pages) available through the Internet. However, many sites do have subscription-based services or areas on their Web sites which require registration and limit access to certain people, for example, veterinarians or members of the organization who maintain the Web page (Figure 4-3).

Why would a veterinary hospital or practice use a computer network? Common uses include:

- *File sharing.* The most common use of a computer network is file sharing. All files that several people may need access to can be stored in a single location and pulled up as needed. This feature is especially useful when a change is made. For example, if you update a document and save it, everyone from that point on will have the most up-to-date version, whereas in the past, each individual person would have to make the same update (Figure 4-4).
- *Printer sharing.* Most offices do not have separate printers for every computer, so a network can be used to share a printer among various users. It can also be used to enable different types of printers. For example, you might print a document on a laser printer in black and white and a graph or color-coded chart on a color inkjet.

FIGURE 4-3 Some professional Web sites may require registration and limit access to veterinarians.

FIGURE 4-4 If entered into the computer, this data would be quickly available to all those who needed it.

- *Sharing database access.* A **database** is a special type of file that contains uniform pieces of information organized into an accessible system. For example, your practice might have a client database with individual fields for the client's name, the patient's name, an address, phone number, and other information. Because databases are often updated frequently, shared access allows each user to have the most up-to-date information.

- *File transfers.* In the past, when you had a file on your computer that you wanted to give to someone else, you had to put it on a disk and transfer it physically. Many files are too large for disks, and you might not have access to a USB flash drive or a CD-writer (CD-R). By using a computer network, you can directly copy a file from your computer to another person's workstation much faster and easier than any type of physical disk transfer.

- *Communication.* An intranet can also be used to facilitate communication between veterinary employees, especially in larger hospitals or offices and when each person has his or her own workstation or computer account. Telephone messages, memos, and quick notes (e.g., "Dr. Peterson, the lab results are back and ready for your review.") can be sent via the network in the form of e-mail or message board directed to the recipient. Such notes are more reliable than pieces of paper because they will not get lost or accidentally thrown away, and only the intended recipient will see them.

- *Sharing Internet access.* If your practice is large enough to have a computer network, it most likely also has a **broadband** Internet connection, which is a connection that is constantly on and, with the help of the network, can be shared by any or all computer users at the same time. Broadband connection has pretty much replaced the use of a dial-up connection via the telephone and the connection through broadband is much quicker.

The Veterinary Practice Web Site

It is becoming more and more common for veterinary practices and hospitals to have an Internet presence in the form of a **Web site**. Like most other business Web sites, it contains basic information such as the name of the practice, the address and phone number, office hours, a brief description of the practice, and the names and credentials of the veterinarians and other staff. Common components of a veterinary Web site might include:

- driving directions and a map
- an e-mail link so that visitors may ask questions
- links to related sites, such as general animal-interest sites, information about particular breeds, veterinary news sites
- an online library of articles, definitions, common ailments
- online appointment scheduling, which would involve clients entering information into a form and requesting a certain day and time to come in, and then having a member of the staff contact them to verify or change the appointment

- prescription refill information and online request forms
- listing of specialty services (such as radiology, laser surgery, veterinary dentistry) and explanations of what is involved
- general pet health and care information
- practice newsletter which may be used to promote special discounts, clinics, or other events, such as National Veterinary Technician Week or a heartworm clinic in the spring
- employment opportunities

If your practice has a Web site, you might be assigned the responsibility of checking and answering client e-mails and contacting clients confirming online appointment and prescription refill requests. If so, you need to incorporate these duties into your daily routine, checking the e-mail each morning when you come in, and at designated points throughout the day. You should return e-mails and other online communications in as timely a manner as you would respond to a telephone call or letter. Responding quickly and effectively to online requests reflects positively on the practice and shows that the staff are caring and professional.

SUMMARY

Computers have become an important part of veterinary practice management and it is essential that all staff learn how to use them. The physical components of a computer are called *hardware*, and the programs used on the computer are called *software*. There are a number of different types of computer programs that are used for different purposes, including word processing, database management, e-mail and Web browsing, as well as specialized veterinary software used for medical record organization, scheduling, and multiple other functions from inventory control to staff management.

Interconnected computers are called a *network*. Most practices that use computer networks use an intranet, which allows users to share information and files, communicate on line, and share printers and Internet access. Your practice might also have a Web site, which may allow clients to request prescription refills and schedule appointments on line. If so, you might be responsible for keeping track of such online requests and communicating with clients in a timely and professional manner.

SCENARIO

MJ is a veterinary assistant employed as a receptionist for a busy animal clinic. One of her daily responsibilities is to check the e-mail each morning to see if any clients have sent in questions, prescription refill requests, or to schedule an appointment. She jots down the names of two clients requesting prescription refills and sets the note aside to pass along to the veterinary technician who will fill the prescription if approved by the veterinarian. She also sees that Mr. A sent in a request at 10:30 the previous morning, asking to have an appointment at 2:00 the next afternoon. "Too bad," she thought, "If he had only e-mailed earlier, I would have been able to fit him in." Unfortunately Mrs. B called yesterday afternoon asking to come in at the same time, and the appointment was given to her. MJ replied to Mr. A's e-mail sending a list of available times, and then logged off.

- How did MJ use the computer to aid in her work?
- How could MJ have prevented the scheduling conflict?
- How else could MJ have used the computer to make her job go easier and faster?
- How should MJ use the computer again today? Why?

REVIEW

Indicate whether the following statements are true or false. If false, indicate how the statement could be made true.

1. The best way to learn how to use a computer is to read.
2. The computer monitor is considered a piece of software.
3. The computer keyboard is a type of hardware.
4. A laser printer uses toner instead of a ribbon.
5. Microsoft Windows is a popular word processing program.
6. If a veterinary practice has a computer network, it will most likely be an intranet.
7. Veterinary computer programs can only be used to diagnose medical conditions.
8. Broadband connections are faster than dial-up connections.
9. A computer network can be used to share printers and files.
10. A Web site can be used both to provide information to clients and to facilitate communication.

ONLINE RESOURCES

Webopedia

This site is an online dictionary of computer and Internet terms. Visitors may search by keyword or category, review the latest additions, the top 15 terms, or the "term of the day."

<http://www.webopedia.com>

MyVetOnline Directory of Veterinary Web sites

Use this site to locate veterinary practice Web sites all over the country.

<http://www.myvetonline.com>

Microsoft

Microsoft is the manufacturer of the Windows operating systems and the Microsoft Office suite of applications. This site features information regarding the latest releases, free updates, and driver downloads.

<http://www.microsoft.com>

Apple

Apple is the manufacturer of the Macintosh series of computers and operating systems. This site features information regarding the latest releases, free updates, and driver downloads.

<http://www.apple.com>

Corel

Corel is the manufacturer of the WordPerfect suite of office applications. This site features information regarding the latest releases, free updates, and driver downloads.

http://www.corel.com

Spectrum Management Systems

Software designed specifically for veterinary practices.

<http://www.vetsystems.com>

Veterinary Practice Management System

Veterinary Practice Management System provides links to VCA Antec, Abaxis, Heska, and Idexx

http://www.viainfosys.com

AVI Mark Veterinary Management Software Systems

This software system offers "Webinars," monthly tutorials on various topics to increase user knowledge of the system. It also has blogs, a Yahoo group, and AviMark forum for user discussions. It is integrated with all VetScan laboratory equipments.

<http://www.Avimark.net>

PART 2

Communication and Professional Growth

CHAPTER 5

Interpersonal Communication

OBJECTIVES

When you complete this chapter, you should be able to:

- describe the communication process and list the basic elements of communication
- list and explain five personality traits that are essential for interpersonal relations in the veterinary office
- distinguish between appropriate and inappropriate professional interactions with clients
- recognize prejudice, insensitivity, or discrimination in interpersonal relations
- list and describe at least five barriers to effective communication
- list three ways you can improve your speech

KEY TERMS

communicate

interpersonal communication
  skills

communication process

message

feedback

reference points

relate

patience

kindness

tact

courtesy

empathy

listening

body language

paraphrase

verbatim

prejudice

stereotype

volume

pitch

tone

monotone

enunciation

Introduction

We **communicate** with each other all the time by sharing thoughts with spoken words, written words, and body language; we exchange information by communication. However, not all of our communication is effective. For instance, did you ever try to have a conversation with someone who for some reason or another was not listening to you? Or have you ever tried to read a letter that was too difficult or written too poorly for you to understand? It is frustrating when the lines of communication are not completely open.

As a member of the office management team, you will rely heavily on your **interpersonal communication skills**, that is, the way you communicate with other people. In your new job, you will need to communicate successfully with office staff and clients. You will handle incoming and outgoing telephone calls, schedule appointments, and greet clients. In order to be successful, you need to develop and understand good interpersonal communication skills.

In this chapter, you will learn how to communicate verbally with clients, both in the office and over the telephone. You will also learn how to interpret nonverbal communication—body language, eye contact, appearance—so you can make sure the nonverbal messages you are sending to clients are positive. However, before you learn about interpersonal communication, it is important to understand how communication works.

The Communication Process

The **communication process** has four essential elements:

1. message
2. sender
3. channel
4. receiver

A **message** is an idea that one person (the sender) wants to get across to another person (the receiver). The person receives this message through a particular channel of communication whether it be written, spoken, or even a meaningful glance. This is known as the communication process (Figure 5-1). The receiver can send a return message, which is called **feedback**. The whole process is then reversed—the receiver becomes the sender, and the sender becomes the receiver.

Even though it sounds like a simple process many things can interfere with a person's message, for example, all senders and receivers have their own **reference points** that determine how they express and understand messages. Education, experience, social and cultural barriers, and religious beliefs are some of the most significant reference points in our lives. So, even though you think your message is clear, a client may interpret what you have said differently because of his or her reference points.

Other factors can interfere with effective communication. Outside noises or actions may distract both the sender and the receiver. In writing, the author's

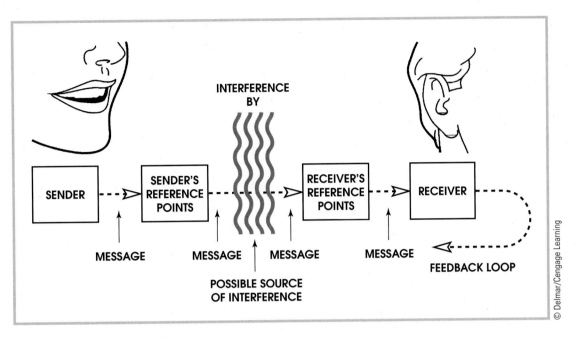

INTERFERENCE
BY

SENDER

SENDER'S
REFERENCE
POINTS

RECEIVER'S
REFERENCE
POINTS

RECEIVER

MESSAGE   MESSAGE   MESSAGE   MESSAGE   FEEDBACK LOOP

POSSIBLE SOURCE
OF INTERFERENCE

© Delmar/Cengage Learning

FIGURE 5-1   The communication process.

style may be confusing, the handwriting may be poor, or the reader may have difficulty reading and comprehending the message.

Good communicators adapt their speech to the listener's needs, expectations, and ability to comprehend. In your position, you must be able to "read" other people quickly so you can send them messages they understand and appreciate. For example, some clients will have little or no knowledge of veterinary medical terminology, so you should always define any terms you need to use when you speak with them or better yet, avoid using specific medical terms and jargon altogether. By contrast, other clients may be very well aware and understand a great deal about their pets' health and medical issues. These clients might be offended if they perceive that you are talking down to them.

Traits for Positive Interpersonal Relations

Think of someone you enjoy talking to. Chances are the two of you have some things in common, that you **relate** to each other in some way. It is just as important that you should try to relate to clients in the same way. When you relate to someone, you make a connection with him. Relating to people is not difficult; you already share some things in common, your concern and love for animals. You will relate to some of your clients without even trying. With others, however, you may have to work at it a bit. Developing the following personality traits will help you relate to all of your clients:

- patience
- tact
- kindness
- courtesy
- empathy

To have **patience** means to deal with difficult situations calmly and without complaint, annoyance, or losing your temper. In a veterinary office, you have to be patient with animals—and their owners. Not all of your clients behave as you would like them to. Some clients continually show up late for appointments. Others bring their pets into the office unrestrained, even though you may have asked them repeatedly to use a leash. Once in a while, clients will even blame you for things beyond your control, such as the length of time they have to wait to see the veterinarian or the amount of their bill. You need great patience to deal with people like this, and it is important for you to resist the temptation to react. Consider the following examples:

> *Impatient*: I'm sorry, but it is not my fault that an emergency arrived just before your appointment and you have to wait a bit.

> *Patient*: I'm sorry that the emergency has caused a delay for you. I know you are busy too. Would you like to wait or would you like to reschedule the appointment? First, let me just check to see how long the wait is likely to be.

Also, you need to resist the temptation to hurry clients, to get them in and out as quickly as possible. Each visit needs to be a positive, welcoming experience for them, not leaving them feeling like they have been in a rushed assembly line. It also helps to offer some refreshment such as coffee, tea, or even a glass of water.

Kindness means being helpful, compassionate, and friendly. It means treating others as you would want to be treated if the situation were reversed.

> *Insensitive*: Hi, have a seat.

> *Kind*: Good morning. It's nice to see you and your little friend again. How have you been? (Pause for response from client) Let's just get your file updated, a quick weight check and the doctor will be ready for you.

Tact means doing and saying the right things at the right time. If you are tactful, you are diplomatic and skilled in dealing with potentially sensitive issues. You can maintain good relations with others and avoid offense. Sometimes, to avoid potential communication problems a nonverbal cue such as a posted sign or directive can eliminate misunderstandings that could lead to the potential of verbal conflicts. Simple courtesy signs such as the one pictured in Figure 5-2 can nonverbally communicate your requests. Front office staff need to be perceptive and understand their own feelings and those of others. Oftentimes, it is not *what* is said but *how* it is said that causes offense. For instance, a client is very late for an appointment:

> *Tactless*: Your appointment was 45 minutes ago. Now you'll just have to wait until we can fit you in. It may be a while.

> *Tactful* friendly greeting: Hello. We were getting a little concerned. Is everything alright? (The client has a chance to respond and reports that the roads are terrible and there was a major accident on the main street.) Why don't you have a seat for a few minutes. Can I get anything for you? We will try and get you in just as soon as possible.

Photograph by Kathy Nuttall, courtesy of Riverton Veterinary Hospital, Riverton, UT

FIGURE 5-2   Not all communication is verbal. Here, a sign requests that all dogs should be leashed. Also note that for client compliance, leashes are available.

You also need to be tactful when you tell clients that their animals or their children are disrupting the reception area.

> *Tactless*: Will you please stop your dog from barking!? He's so loud that I cannot hear the person on the telephone.

> *Tactful*: Excuse me, but Rex is so excited I have a problem hearing the person on the phone. It will be about 10 minutes yet for the doctor to see you. Why not take Rex outside for a little bit and help him settle down.

> *Tactless*: Kids, stop running around and making all that noise.

> *Tactful*: Children, would you like to have some magazines and coloring sheets? You can use that little table in the corner.

Courtesy means putting the needs of other people before your own. It means cooperating, sharing, and giving. You should treat all clients on a polite, professional, and impartial basis. *Please, thank you, you're welcome, excuse me,* and *how may I help you?* should become standard phrases in your vocabulary. You must be careful not to play favorites; do not do special favors for one client that you would not do for another or make a big fuss over one pet and ignore the others. To the owner, every pet is special.

Empathy means being able to feel and understand what another person is feeling. When you empathize with your clients, you show that you understand what they are feeling.

CLIENT: I've never felt so exhausted in my life. Housebreaking a puppy is hard work.

RECEPTIONIST: House-training can be very tiring, being on your toes all the time to prevent accidents. Try this; it is a very helpful brochure with special tips on house-training. Maybe this can help make it a little easier.

Interacting With Clients and Co-Workers

Speaking with Clients

During an average day, veterinary staff could answer more than a hundred questions. Front office staff spend much of the day answering clients' questions and giving them advice about pet care. Sometimes, it can be frustrating. Clients will ask you the very questions that you have just finished answering. Other times, you may feel clients' questions are irrelevant, unnecessary, or a waste of time. But no matter what the question is, you should never appear rushed or irritated. Answer all questions patiently and tactfully.

One way to help clients with their questions is to give them written instructions along with your verbal instructions.

STAFF: The name of the drug and directions for giving these tablets are right here, on the bottle: one pill, with food, three times a day. It helps to use a special treat when you give the medicine. Would you like me to write the instructions on your receipt as well? And don't hesitate to call if you have any questions at all.

A good technique to use is the "echo" technique to make sure clients understand directions. To do this, ask the client nicely to repeat the directions back to you "to make sure there are no misunderstandings." Then, listen very carefully to make sure the instructions have been clearly understood.

STAFF: Do you have any questions about how and when to give the medication?

CLIENT: You said she gets three a day, with food. Right?

STAFF: Perfect.

Because you work in an animal hospital, clients will often ask your advice about their pet's medical problems. You must never discuss different types of treatments with clients, suggest a diagnosis or possible outcome. Never interject your own personal experiences with your own pet's problems or those of a similar case. Tactfully advise the client to discuss all concerns with the veterinarian.

CLIENT: Do you think this lump on Millie's head could be cancer?

STAFF: I'm sorry, I really don't know. I understand that you are anxious but the doctor will examine it and help you with your questions.

CLIENT: If it is cancer, do you think there is a cure?

STAFF: I'm sure the doctor will be able to answer all of your questions once she has made a diagnosis.

When you are interacting with clients, make sure your feedback is appropriate. That is, the feedback you give should be a true reflection of your concern and understanding of the message. For example, if the staff member in the previous example had snapped her response in a harsh voice, the client would sense impatience and annoyance behind the words. However, if she used a pleasant voice, made eye contact, and punctuated her words with a comforting smile, the client would sense the genuine care and concern that veterinary staff should be projecting toward clients. The staff member shown in Figure 5-3 is projecting a positive, upbeat image to the client.

Listening

Picture this scene: You have just shown a client into an exam room and when you turn to leave , see that she is leaning over her 12-year-old Irish setter holding her head in her hands. Concerned, you ask, "Are you okay? Can I get you something?" The client looks up, but does not look you in the eye when she answers, "Yes, I'm fine. I'm just a little bit tired." However, when you look at her, you see that she is crying. She tells you with words that she is fine, but the rest of her says she is very upset. The client is sending mixed messages that can put you at a loss for words.

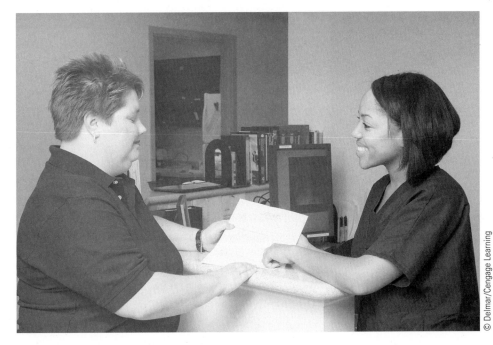

FIGURE 5-3 Eye contact and a smile are strong interpersonal communication tools.

Listening to what others say is an important part of the communication process but listening is not complete without observation. Keep in mind that listening involves the eyes as well as the ears. Good listeners hear exactly what another person says, and they compare that message with the person's expressions and **body language**, the gestures and mannerisms that help a person communicate.

Good listeners also know when *not* to speak. Have you ever had trouble putting your ideas into words, only to be frustrated all the more by someone else jumping in and filling up the silence during which you were thinking? Use short periods of silence to allow clients time to collect their thoughts. Look for nonverbal cues if the silence has gone on too long and is causing anxiety. If a person is fidgeting or rubbing his or her temples, it is time to break the silence.

HOW TO LISTEN

- Prepare yourself to listen. Stop all other activity and focus your attention on what is being said.
- Make eye contact with the speaker.
- Concentrate on what is being said.
- Do not allow distractions of any kind to interfere with your listening.
- Listen with empathy.
- Listen not only to what is being said but also to how something is being said.

Silence at inappropriate moments can sometimes be a sign of hostility or mental disturbance. Once in a while, you may meet a client who is so upset about his or her pet's illness or death that he will only speak to the veterinarian. In this type of situation, you need to know when not to push things (Figure 5-4).

In communication, listening is generally more important than talking.
—*From* The Job Survival Instruction Book

Observing

As you just learned, you need to pay attention to people's body language as well as their words. In order to be a good communicator, you need to be able to recognize nonverbal cues—your own and those of other people. Observing nonverbal behaviors is not always easy. In fact, interpreting nonverbal cues can be complicated.

Often, people do not even realize they are communicating nonverbally. We can control our body language to some degree, but most nonverbal communication is not always under conscious control. Many of us have habits—nail biting, finger tapping, hair twisting—that reveal our nervousness or boredom. Nervous habits are easy for an observer to recognize, but many other types of

© Delmar/Cengage Learning

FIGURE 5-4 Understanding body language is an important part of communication. The client clearly does not want to speak with this member of staff.

nonverbal cues are not so obvious. In fact, the same gesture or facial expression may mean one of several different things. For example, if you sit with your arms folded over your chest, it could mean you are:

- trying to protect yourself from somebody or something
- hugging yourself as a form of comfort
- conscious about your physical appearance
- cold and trying to warm up

Also, keep in mind that nonverbal behaviors do not mean the same thing to everyone. For instance, in North America a nod of the head means yes and a shake of the head means no. But in some cultures, a nod means no and a shake means yes.

You should not jump to any conclusions about the meaning of any particular nonverbal communication. If you are dealing with a client and wish to understand a certain nonverbal message, you might discuss your observations with him. Ask the client about his or her feelings. Just remember that your observation should never be stated rudely or seem judgmental. Use tact, by doing and saying what is right and appropriate at the time. Do not play the amateur psychiatrist or invade the client's privacy.

Rude: You do not have to cross your arms over your chest. The veterinarian will not hurt your puppy.

Tactful: Are you comfortable? I would be happy to turn up the heat if you are cold.

Observing and interpreting nonverbal behavior is especially important when the individual's body language contradicts his or her words. In these cases, the individual may be expressing through nonverbal communication what he is unwilling or afraid to say out loud. For instance, look at the exaggerated nonverbal behavior in Figure 5-5. What is the client's body language saying? How does the nonverbal communication contradict the client's words? Despite his words to the contrary, it is clear from his body language that the client in Figure 5-5 is reluctant to hand his cat over to the staff member, even though he says he will. In this instance, she should try to reassure the client that his cat will be just fine with her. Her words and her actions—removing the barrier of the front desk by coming around to the client, and giving the cat some extra attention—will most likely ease the client's fears.

Paraphrasing

Really good listeners sometimes **paraphrase** what they hear.

Paraphrasing, or using different words to express the same idea, is a great way to develop listening skills. When you paraphrase, you listen to a speaker and then repeat his or her message in your own words without changing the meaning.

FIGURE 5-5 An extreme example of body language. The client's body language does not agree with what he is saying.

SPEAKER: I work so hard all year long. It doesn't seem that I'd be out of line to expect decent accommodations and good weather for two lousy weeks!

LISTENER: Sure, it's part of what you're looking forward to on a vacation and you want to be able to enjoy all of it, including good weather and a nice place to stay.

Repeating the speaker's words **verbatim** (word for word) does not have the same effect. Word-for-word repetition could just mean that you are only repeating what you hear and don't really care or understand what you heard. Besides, clients are apt to get annoyed with you if you are always saying exactly what they say. But saying what they *mean* is a different matter. When you do this, clients feel that you are really paying attention to them and that you are fully engaged with them and what they have to say.

In addition, by paraphrasing, you can test whether you heard the message correctly and understand the speaker's intentions. Hearing their own words reflected back to them by a member of staff also helps clients. People do not always say exactly what they mean. If you missed the point, they can tell by your paraphrasing and then clarify what they really meant to say.

VETERINARY ASSISTANT: Could you describe Muffin's problems for me?

CLIENT: She isn't eating, and she's throwing up a lot

VETERINARY ASSISTANT: So Muffin doesn't want to eat, and she vomited this morning?

CLIENT: Well, actually, she threw up yesterday and she hasn't eaten since then.

Paraphrasing can also enhance a client's willingness to talk. Consider the scenario with the client and her 12-year-old Irish setter. Remember that she has just told you she's tired, even though she's obviously upset.

YOU: Are you sure you're all right?

CLIENT: Well, now that you mention it, I am really worried about Daisy. She's supposed to have an operation, and I'm afraid she might not make it. You know, she's getting really old.

YOU: I can certainly understand your concern but try not to worry. The doctor will take very good care of Daisy. Are there any questions I can help you with or would you like to have a little time with the doctor before leaving Daisy with us? I promise, we will call you just as soon as Daisy is out of surgery.

By paraphrasing the client's response, you were able to get her to open up and discuss what was really bothering her—her fear of losing her dog. Just with this extra compassion, the client was comforted and reassured that Daisy was in caring hands.

Showing Respect

As a member of the office management team, you will be dealing with clients and co-workers of different cultures and races. Treat people equally, as individuals, no matter what personal feelings you have about their social class, skin color, sexual orientation, physical challenge, or any other differences from yourself.

Occasionally, you might have to communicate with a client from another country. If this is the case, he might have trouble understanding the language and culture of your country. This is one reason why it is so important for staff to have empathy, the ability to understand the way others feel, and to respect every human being.

Clients with disabilities may present the need for improving your professional skill, care, and judgment. It is important to remember in dealing with these clients that although they may *have* a handicap, *they* are not a handicap. They are total persons, just the same as you or anyone else.

You will likely come in contact with clients who may be blind, deaf, or physically disabled. Some might have certain mental or emotional disadvantages that other people possess and take for granted. You should never judge, label, or stereotype clients. If you feel awkward around them, you will make them feel uncomfortable. Treat all clients as people first. You should certainly make allowances for whatever disabilities they have, but you should never identify these clients only by their handicaps or disabilities. If you do, you will be gearing your entire approach to them based on what they lack rather than on what they have.

TIPS FOR TREATING CLIENTS WELL

- Always offer a pleasant greeting when clients come into the practice and make eye contact as you greet them.
- Treat all clients equally, with respect and honesty.
- Address clients by their names and title: for example, use Mrs., Ms, Mr., Professor, Doctor, etc., to preface their last names. Use first names only for children or adults who ask you to.
- Try to relieve clients' fears regarding their pets' care by carefully explaining all procedures to them.
- Explain medical terms if it seems that the client does not understand what you are saying. Many people are reluctant to ask what a word or phrase means.
- Be sure that the reception area is kept tidy and comfortable for clients. If any patients have an "accident," clean it promptly and cheerfully to prevent the client from being embarrassed.
- Offer assistance if you observe a client experiencing difficulty with a door, chair, or the animal's carrier or any other situation which may arise.

Respecting others requires you to understand your own beliefs and **prejudices**. Most people do not believe themselves to be bigoted or prejudiced. But in reality, every person is because we all have preconceived opinions and biases. We are all partial to some things and not to others. It can be difficult to recognize our own opinions and biases as prejudices because they seem normal to us. Because of this, most people do not even realize how their personal prejudices affect the way they treat others. Waiters and waitresses often place the bill in front of a man without asking who will be paying. People often call a veterinarian "he" without even thinking she might be a woman. Clients are often expected to be Caucasian if their names do not fit what we might think are ethnic stereotypes.

Stereotypes are preconceived ideas about a group of people made without taking individual differences into account. Stereotyping lumps groups of people together, assigning them the same traits and behaviors, simply because they belong to a certain social or ethnic group. Stereotyping is a type of prejudice. People who really respect individual differences and who really care about other people—all people—do not label individuals. You must be sure that your personal beliefs or prejudices do not interfere with your ability to provide equal, objective professional care to each and every client. If you learn all you can about the cultural groups in your community, it will help you avoid offending people who have different cultural practices than you do, and you will earn their respect and confidence in the process. This does not mean you should question clients about their beliefs or something you may have heard about different cultural events, but to be open and listen and participate in your community.

PREJUDICES THAT AFFECT CLIENT/STAFF RELATIONSHIPS

- Racial or ethnic prejudices
- Gender prejudices (discrimination against women; "man bashing")
- Religious prejudices
- Prejudice against those who do not speak English or those who speak it poorly
- Perceived sexual orientation
- Hidden prejudices:
  - dislike of children
  - dislike of the elderly
  - dislike of overweight people
  - dislike of people on public assistance
  - dislike of the mentally or physically challenged
  - dislike of people with tattoos and/or body piercings

Telephone Manners

It is very easy to forget that the voice on the other end of the telephone line is an actual person, and as such is due the proper attention and respect due to any client. Here are a few points to remember when speaking on the phone:

- *Answering.* When the phone rings, be sure to answer it in a reasonable amount of time. Generally speaking, you should answer before the third ring.
- *Greeting.* Always use an appropriate greeting when answering the telephone. You should identify the name of the practice or veterinarian, give your name, and then ask how you may help the caller. This ensures

the person calling that he has dialed the correct number, explains who he is talking to, and invites him to make his or her request. Refer to the staff manual for the preferred method of answering the telephone for your clinic or hospital.

- *Listen.* Just as you would with a person you were speaking to face-to-face, you should be sure to listen to what the caller is saying, and answer promptly and courteously. Do not do other things while speaking with someone on the telephone. It can be very distracting to you and to the caller. This not only includes trying to attend to another client at the front desk at the same time, but eating, drinking, or chewing gum.
- *Manners.* Use the same manners on the telephone that you would use in person. "Please," "thank you," and "you're welcome" show respect for the person you are speaking with and portray you, and by extension, the practice, as professional.
- *Avoid putting callers on hold.* When a caller is put on hold, he can quickly feel ignored, even though you might be looking up the information he is requesting. Never "rattle off" something like "Veterinarycliniccanyouholdplease"- clunk- without giving the caller to, at the very least, state the reason for the call. It could be an emergency!
- *Take notes.* Keep a pad by the telephone to take notes when anyone calls. Write down the client's name, telephone number, and the nature of the call. This helps reduce the number of times you need to ask the person to repeat the information, and shows that you are paying attention to the call. There may also be the occasion when the call is cut off or the client simply hangs up. It is professional to return the lost call immediately with an apology.
- *Know why you are calling.* If you are the person placing the call, know the reason you are making the call and what you hope to gain from it. Be sure to have any information or files that you might need on hand while making the call to avoid delays. You do not want to waste the time of the person you are calling any more than you would want a caller to waste your time.

Your Body Language

You already know that you need to observe your clients' nonverbal messages in order to understand what they are feeling, but, what about your own nonverbal communication? You are also sending out messages to clients. You need to be aware of your nonverbal messages, and make sure they are positive. *The way you dress, your facial expressions, your eye contact, and your body movements all say something about you and your relationship to the client and to the practice.*

Earlier, you learned how important it is to convey a professional image. When you take pride in your appearance, you send clients a message that you are a professional who is in control and willing to be of service to them and their pets. Always practice good hygiene, groom yourself attractively, and dress appropriately.

Because veterinary assistants work with animals, they usually dress casually. Keep in mind, though, that first impressions are important, so "casual" does not mean "sloppy." Almost all veterinary clinics and hospitals provide scrub tops or polo shirts with the clinic logo. Use these and make sure they are spotless. Avoid wearing a lot of jewelry which can present a safety hazard to both you and to the patient. A sensitive issue is that of visible tattoos. Many people find tattoos in poor taste and have certain judgmental or preconceived ideas about people with tattoos, just as discussed with the other issues of prejudices. This concern should be addressed with an employee by the practice manager should it ever arise.

Another obvious positive nonverbal message is a smile. All members of the veterinary team should be quick to smile when interacting with clients. A comforting, sincere smile can go a long way toward relieving anxieties. However, be careful not to force a smile; you do not want clients to feel you are faking your hospitality. When a client is deeply troubled, be especially careful to smile only at appropriate moments so that you will not be perceived as insensitive or flippant. Your facial expression should always be appropriate to the tone of the message you are sending or receiving. But do not frown. If you look glum, you will only add to the nervousness of clients whose animals are sick or in pain.

> **GOOD BODY LANGUAGE FOR THE VETERINARY OFFICE ASSISTANT**
>
> - Face the client and really look at them, maintaining friendly eye contact.
> - Hold your arms relaxed at your sides or place them on the desk or counter.
> - Stand straight, keep your posture erect but not rigid.
> - Stay approximately one arm's length away from the client.
> - Always wear professional, clean attire.
> - Always practice good hygiene.
> - Keep a relaxed, open facial expression.
> - Speak in a moderate and clear tone of voice.
> - Wear a name tag.

Eye contact is a very important nonverbal method of communication. When you look someone in the eye, in effect, you are saying, "I am interested in speaking with you and hearing what you have to say." Do not look away from people when they are talking to you, or they will feel that you are not interested in them. But do not stare intently, either, or you will communicate the opposite—too intense interest, or even hostility.

If you make eye contact when you speak, clients will sense that you are open and honest. If you do not make eye contact with clients, they may perceive you as being uncertain of what you are doing and that what you have to say may not be valid or trustworthy information. Eye contact gives you a good idea of what might be going on inside someone's head. You can usually tell whether clients have understood your messages by the look in their eyes.

Your gestures (body movements) can also enhance your interactions with clients. A welcoming or parting handshake is a powerful gesture that is a sign of friendship. A comforting touch on the shoulder or a pat on the back at an appropriate moment can make a client feel secure. However, some people do not like to be touched by strangers, so do not approach every client in the same way. The touch has to be a genuine, friendly gesture attached to a moment of closeness. For instance, you might put your hand on the shoulder of a child whose dog is sick or hurt.

Handling Angry Clients

Occasionally a client may become angry. More often than not, this anger is directed at front office staff. Frequently, the attack comes so suddenly that you feel shocked, assaulted, and frustrated. The situation will be easier to deal with if you realize that the person is in a highly emotional state and the anger is not directed at you personally.

Unfortunately, there is no magical formula for calming people down. If you know the person very well, you might be able to figure out the source of anger and then deal with it. The real source of anger may be different from whatever subject the person is yelling about. However, you will seldom know clients well enough to do anything except listen and that is a very important first step in trying to resolve the problem.

Keep a positive attitude; that is, try to respond to people, situations, or objects in positive rather than negative ways. That goes for your nonverbal messages too: make them positive. Remember, a frown shows disapproval rather than empathy. The following list indicates other negative nonverbal behaviors:

- A mumbling tone of voice shows lack of respect.
- Fidgeting or turning away from the client shows a lack of warmth or a desire to escape.
- Evasive eye contact indicates insincerity.
- Pointing a finger, shaking your fist, or speaking in a loud voice sends a confrontational message.

Instead of these negative nonverbal behaviors, use positive head nods, devote your full attention to the speaker, paraphrase what is being said, be honest, look the speaker in the eye, and use a natural tone of voice.

Here are some additional tips to help you cope with an angry client:

- When you notice that a client is becoming angry, show him or her into a private area such as an examining room, an office, or some other less-public spot.
- Permit the client to tell you about the problem. Remain objective. Listen attentively.
- At no time should you place any blame for the problem or accept responsibility for any situation. Do not say anything critical of the veterinarian or any other staff member. Do not say anything the client might interpret as being critical.

- Keep yourself under control, no matter what the angry client says. At all costs, avoid arguing.
- Apologize for any misunderstanding.
- Assure the client that the matter will be resolved and that any conflict will be avoided in the future.

Angry people often say things they regret later, so do not take personal offense at anything an angry client says. If the client raises a reasonable objection to an error made by someone in the office, do not get defensive and try to justify the error. The client's complaint will be handled by the veterinarian, so you should say nothing to indicate that the conflict will be resolved in any certain way. If you treat an angry client with tact and understanding, you could save the veterinarian's reputation from the damage a disgruntled client can cause.

Anytime you argue with a customer, you lose. Even if you win, you lose.
—From The Job Survival Instruction Book

Veterinary Team Communication

By now, you know you need good interpersonal communication skills to communicate with clients. You also need to use your interpersonal communication skills when you interact with the veterinarian and your co-workers. The veterinary staff is a team working together. Although you will probably be assigned specific duties, keep in mind that you are a team member and you need to be flexible. *"That is not my job" is one expression that you should eliminate from your vocabulary*. You also must be careful not to overstep your boundaries. There is a difference between pitching in and taking over.

As you can see from the employee evaluation form in Figure 5-6, the ability to cooperate and get along with others is very important to employers. You need to be cooperative, dependable, polite, and patient with your co-workers, even though your co-workers may differ from you in their personality traits, beliefs, values, and work habits. You must be above pettiness and moodiness. Everyone gets down in the dumps or irritated sometimes, but the competent and professional staff member does not bring a bad mood into the office. Never let your personal problems interfere with your professional interactions, even when your problem is with a co-worker. Put the objectives of your job before your personal feelings.

The veterinarian is the leader of the veterinary team. If you work for a small private practice, you may be the only assistant and perform both office management and clinical duties. In this case, a good relationship with your employer is even more important than in larger veterinary clinics and animal hospitals. But regardless of where you work, a good working relationship with the veterinarian is always important to a winning veterinary team.

Employees who are helpful and easy to get along with are valued more than difficult people with better skills.
—From The Job Survival Instruction Book

Employee Evaluation

Date of Evaluation:

Evaluator:

Employee Name:

Date Hired:

Position:

How Long in Present Position:

Supervisor:

Employee's Job Description:

Employee Performance Review

The object of this evaluation is to assess the employee's performance in his current position and to assess his commitment to. This evaluation should highlight the employee's strengths and successes and identify areas for improvement.

In each of these sections the employee is measured on a numerical scale from 1 to 5 as follows:

Consistently exceeds expectations	5
Frequently exceeds expectations	4
Meets and/or occasionally exceeds expectations	3
Occasionally does not meet expectations	2
Frequently fails to meet expectations	1

This evaluation is divided into the following sections:

1. Communication
2. Personal/Motivation
3. Interpersonal Skills
4. Knowledge/Skills
5. Career Development

Section 1: Communication

_____ consistently meets expectations in this category.

Comments by Evaluator:

_____

FIGURE 5-6   A sample of a good employee evaluation form, which also provides space for an employee's comments.

(continued)

Section 2: Personal/Motivation

_____ consistently meets expectations in this category.

Comments by Evaluator:

_____

Section 3: Interpersonal Skills

_____ consistently meets expectations in this category.

Comments by Evaluator:

_____

Section 4: Knowledge/Skills

_____ consistently meets expectations in this category.

Comments by Evaluator:

_____

Section 5: Career Development

_____ consistently meets expectations in this category.

Comments by Evaluator:

_____

Section 6: Overall Rating

20 ☐ 25 points	Superior performance. Exceeds overall expectations.
15 ☐ 19 points	Fully competent. Meets the performance expectations of the job.
0 ☐ 14 points	Development required to meet expectations.

_____ has scored the following:

Communication   3

Personal/Motivation   3

FIGURE 5-6   *(Continued)*

(continued)

Interpersonal Skills   3

Knowledge/Skills   3

Career Development   3

_____ has a total of 15 points. Overall, he is fully competent and meets the performance expectations of the job.

_____

Section 6: Employee Comments

_____

_____

_____

_____

_____

_____

Section 6: Improvement Plan

_____

_____

_____

_____

_____

We have read and discussed this evaluation:

_____   _____

                                                                          Date

FIGURE 5-6   *(Continued)*

Your attitude toward the veterinarian must be respectful. Always call the veterinarian by his or her professional name.

> VETERINARY ASSISTANT: Good morning, Dr. Masterson. It looks like you have a full day.

> VETERINARY ASSISTANT: Please make yourself comfortable, Mr. Levine. Dr. Masterson will be with you in a few moments.

The veterinarians you work for are entitled to your loyalty. They are highly regarded professionals working in an often physically and emotionally stressful position. Although you might not agree with a particular behavior, action, or outcome, you should be careful not to assume the cause is a lack of skill, knowledge, reliability, or professionalism. Veterinarians generally work to the best of their abilities, with the best interests of the client and patient at heart. This does not mean that you or the clients will always agree with a veterinarian's recommendations. However, even if you feel strongly that the veterinarian has done something wrong, you should not express this to clients, co-workers, or any member of the general public. Never say or do anything that would cast an unfavorable light on his or her reputation. If you are experiencing a problem with a veterinarian, speak to your supervisor in private.

Discrimination

In the section "Showing Respect," you learned about discrimination in regard to your clients. You learned how to recognize your own prejudices and insensitivities. But what do you do if the discrimination is directed toward you? By law, employers are not allowed to discriminate based on race, color, national origin, religion, sex, family status, handicaps, or age. If you feel you have been denied a job or have lost a job for any of these reasons, contact the Equal Employment Opportunity Commission (EEOC) or the Canadian Labour Relations Board. However, if you are the victim of a more subtle discrimination—say, a co-worker makes offensive comments about your weight, or you feel a client is harassing you—then you will have to use interpersonal skills to handle the situation.

GUIDELINES FOR VETERINARY TEAM MEMBERS

- Be sensitive to the people you work with.
- Make adjustments to cooperate with fellow employees.
- Show interest.
- Do not take personal problems into the office.
- Express appreciation to teammates.
- Be courteous.
- Be open to new methods and concepts.
- Keep communication lines open between all staff members.
- Be honest with yourself and others.

(continued)

(continued)

> • Keep the private business revealed to you by co-workers to yourself.
> • Do not spread rumors; if you have a criticism of an employee, take it directly to that person.
> • Do not gossip about clients, veterinarians, or other personnel to anyone, not even your spouse or your best friend.
> • If misunderstandings occur, clear them up immediately.
> • Admit your mistakes.
> • Accept constructive criticism graciously.

VETERINARY ASSISTANT: Mary, thank you for your concern, but I have heard enough comments about diet programs and exercise videos.

VETERINARY ASSISTANT: Mr. Stern, as I have told you, I simply do not date clients. Please do not make any more personal advances.

In some situations, confronting the discrimination may be risky business. What if your employer, an employer's relative, or a favored employee is doing the discriminating? In these instances, reporting the situation may result in your termination or even in a messy legal proceeding. You will have to judge for yourself whether to turn your back, resign, ask for a transfer, or confront the situation. Often, you can help correct a situation by saying that you are uncomfortable working in an environment in which unfair practices are going on but you must be committed and be able to back up, with specifics, the issues of concern .Otherwise it becomes just a vague threat which could lead to even great problems. Employers may be sued by employees who are harassed.

Improving Your Speech

A person's speech can interfere with his or her message. For instance, if a person speaks too quickly or mumbles, you might miss some of what was said. If a person speaks too slowly, too softly, and uses poor grammar, you might lose interest.

When you speak to clients and co-workers, try to speak at a moderate rate. Speak clearly and effectively and use direct, concise language that the client can understand.

The use of correct grammar and pronunciation, and speaking in a pleasant tone of voice also contributes to the effectiveness of what you are saying. Analyzing the speed of your voice to see if it needs improvement is a good exercise.

How quickly or slowly do you speak? Do people often ask you to repeat a comment? Do they seem uninterested in what you say? Your voice should sound "natural." If you have a natural tendency to be a fast talker, slow down. If you speak too slowly, speed up. An average rate of speech should be approximately 120 words per minute. You can measure your rate of speed by reading the passage in Figure 5-7 out loud, taking time to pause where you would if

COMMUNICATION BARRIERS

While there are many barriers to effective communication, Thomas Gordon, an expert on interpersonal communication, has identified 12 of the most common ones. These conversation stoppers are almost guaranteed to block the flow of communication between individuals, and can even end friendships! How many do you recognize?

Criticizing. Making a negative evaluation of the other person's actions or attitudes. "You brought it on yourself; you've got nobody else to blame for the mess you're in," or "Can't you do anything right?"

Name calling. Putting down or stereotyping the other person. "You hardhats are all alike," or "What a dope!" or "Just like a woman," or "You're really dumb."

Diagnosing. Analyzing why a person's behaving a certain way; playing amateur psychiatrist. "You're just doing that to irritate me," or "I know just what's wrong with you," or "Just because you went to college, you think you're better than I am."

Praising evaluatively. Being too nice by saying things about a person that are excessive or aren't really true. "You're perfect," or "You're the best typist in the world," or "I've never seen anything like that report, really fabulous."

Ordering. Commanding the other person to do what you want to have done. "I want you to do this report right now. Why? Because I said so!" or "Get these letters out right now and take your break later."

Threatening. Attempting to control the actions of others by warning of negative consequences. "If we don't get along better, I'm going to tell Mr. Smith about you," or "You'll finish that report tonight or else!" or "Just come in late again and see what happens."

Moralizing. Telling another person what to do or preaching what you believe is right or proper. "You shouldn't get a divorce; think about what will happen to the children," or "You ought to tell him you're sorry," or "You can do much better than that if you try."

Bully questioning. Asking questions that are often conversation stoppers because the response must be a forced yes or no. "Are you sorry you did it?" or "Well, weren't you supposed to know that before you attended the meeting?" or "You mean you didn't take the report with you?"

Unwelcome advising. Giving the person a solution to a problem even when the person didn't ask for one. "If I were you, I'd sure tell her off!" or "That's an easy one to solve; first you . . . ," or "What you need to do is go to night school."

Diverting attention. Pushing the other person's problems aside through distraction. "Don't dwell on it, Sarah; let's talk about something more pleasant," or "You think you've got it bad—let me tell you what happened to me!"

Logical argumentation prematurely. Attempting to convince the other person with an appeal to facts or logic without knowing the factors involved. "Look at the facts: if you hadn't left work early the other afternoon, we would have finished the report, and Ms. Smith wouldn't be upset," or "By devoting 20 minutes to opening the mail in the morning and concentrating on getting all your typing done before lunch, you should be able to spend every afternoon changing the files over."

False reassuring. Trying to stop the other person from feeling negative emotions. "Don't worry, it's always darkest before the dawn," or "It will all work out okay in the end," or "There's no point in crying over something that you can't do anything about."

FIGURE 5-7   Read aloud this passage of 600 words to check your rate of speaking.

you were engaged in a conversation. Read the passage through silently once or twice to familiarize yourself with the words. Time yourself. When you finish, divide 600 by the number of minutes it took you to read. Do not round off to the nearest minute, but to make the math easier, round off to the nearest 15 seconds. For instance, if it takes you 5 minutes and 15 seconds, divide 600 by 5.25. If it takes you 5½ minutes, divide by 5.5. If it takes you 5 minutes and 45 seconds, divide by 5.75. A speaker with an average rate of speech will take approximately 5 minutes.

So, how did you do? Do not worry if the rate at which you speak is less than perfect. Remember the old adage, "practice makes perfect."

Volume, Pitch, and Tone

Your speaking **volume** is the degree of loudness. The **pitch** of your voice is its highness or lowness of sound. **Tone** communicates mood or feeling; your voice can sound soft, rough, sweet, harsh, excited, or bored. The volume, pitch, and tone of your voice will vary according to circumstances. Listen to someone who is excited about something. That person's voice will have a high, louder-than-usual quality to it. Or, listen to someone giving a speech over a microphone; the tone will usually be lower and richer.

Some people speak so loudly that they blast the listener's eardrums, while others speak so softly that they can hardly be heard. It is difficult to concentrate on either type of voice. Of course, there are times when shouting and whispering are the appropriate speaking volumes. But do you shout or whisper when you speak in normal conversation?

Although it is good to maintain a moderate volume, pitch, and tone in the office, moderation should not be taken so far that it becomes a monotone. Speaking in a **monotone** voice—one that does not show a change in feeling or pitch—is a quick way to put your listener to sleep. A voice with variety is more pleasant to hear than a constant humming sound. Raise and lower your voice as you speak. This variety makes you appear more interesting and people are more likely to listen to what you have to say. Use a pleasant tone of voice that shows enthusiasm and warmth. Remember that your voice represents your personality. Your speech should match the smile on your face.

Did you ever notice that some people change their voices when they are on the telephone? The phone can bring out the worst in people's speech. Some people who speak at a moderate-level volume in face-to-face conversation will use the telephone like a bullhorn. Others speak as if the phone wires amplified their voices. Since a significant part of the veterinary assistant's job includes answering the phone and making calls, you should have a friend critique your telephone use. Arrange with someone to receive your call and place one to you. Have your friend note your volume, pitch, and tone.

Enunciation and Pronunciation

The way you form or articulate your words is called **enunciation**. To enunciate clearly, you must use your lips, teeth, jaw, and tongue to form precise sounds and not slur words together. To practice your enunciation, read out loud into a tape recorder, concentrating on each word, and play back your recording to hear how you sound. Avoid the following common mistakes:

Do not sound a silent *h*.

- heir
- honor
- heiress
- honest
- honorable

Be sure to sound the *h* in each of these words.

- wharf
- when
- where
- which
- while
- whip
- whiz
- why

Distinguish between the sound of *ern* and the sound of *ren*.

- south*ern*
- west*ern*
- north*ern*
- east*ern*
- child*ren*
- breth*ren*

Do not confuse *pre* with *per*.

- *per*form
- *per*sist
- *per*haps
- *pre*tend
- *pre*vent
- *pre*scription

Sound the final *g*, but do not hang onto it, and do not make it hard like the *g* in *grunt*.

- sitting
- playing
- dancing
- sing
- ring
- thing

Do not run words together.

- Give me (not *gimme*)
- Saw her (not *saw r*)
- Let me (not *lemme*)
- Catch them (not *ketch em*)
- Don't you (not *don't cha*)
- Going to (not *gonna'*)

Other very commonly mispronounced words are "pacifically" for specifically and especially important, "vetrinary" instead of veterinary.

TONGUE TWISTERS

If you practice each of the following tongue twisters, you will improve your pronunciation. Try modulating your pitch and tone in an interesting way.

1. Are our oars here?
2. Bring me some ice, not some mice.
3. Suddenly seaward swept the squall.
4. He saw six, long, slim, slender saplings.
5. Amos Ames, the amiable aeronaut, aided in aerial enterprise at the age of 88.
6. Six thick thistle sticks, six thick thistles stick.
7. A big black bug bit a big black bear.
8. Geese cackle, cattle low, crows caw, cocks crow.

Additional Tips

You might enjoy taking speech, oral communication, or acting classes to improve your oral communication skills. But you can also become more articulate simply by keeping your ears and eyes attuned to language. Here are some suggestions:

- *Listen attentively to those who speak correct, effective English.* Pattern your speech after theirs. You might ask a friend who speaks well to call your attention to any errors you make. Avoid the use of double negatives, such as "you don't want not to feed a good diet."
- *To train your ear, imitate a favorite radio or television announcer.* Be sure to choose an announcer with perfect enunciation. Listen to this announcer whenever possible, paying close attention to his or her speech patterns. Imitating this speech pattern, speak into a tape recorder and play it back so you can analyze your progress.
- *Listen to recordings of popular books.* Tape recordings of books are narrated by excellent speakers, and most books are written in Standard English. Be aware that dialogue in fiction is often written in colloquial or local dialect. Picking out the differences between Standard English and the various dialects is a good way to improve your language skills.

- *Acquire the dictionary habit.* You will hear, and discover in your reading, many unfamiliar words. Do not let new words pass you by. Look them up in the dictionary. Note their spelling, pronunciation, and meaning. Whenever there is a suitable occasion, use them so that they become an active part of your daily vocabulary.

SUMMARY

Effective interpersonal communication is an important aspect of working in the field of veterinary medicine. You have to know how to get your message across clearly and appropriately in a variety of situations. Communication involves listening, speaking, and being aware of body language, that of yours and of those who are speaking with you. You must always be respectful when speaking with clients, in person and on the telephone, even when they are angry. Interpersonal communication skills are also vital to your interactions with the veterinary team. Attentive, respectful interactions make for a stronger and more enjoyable working environment.

The way you speak is often as important as what you are saying. You should always speak clearly, with proper enunciation and correct pronunciation, and ensure that the tone, pitch, and volume of your voice match the message you are trying to convey.

SCENARIO

A client arrived for her cat's appointment with her two rambunctious boys in tow. She checked in, with AD the receptionist, and then took a seat with the cat on her lap while the boys chased each other around the room. Their mother quietly tried telling them to sit still, obviously embarrassed by their behavior, and clutching the cat a bit too tightly as she spoke to them.

The receptionist was not very fond of young children, and wanted to yell at them and tell them to quiet down, but he knew it was not his place. He was also concerned about the way the client was holding her cat, which was now starting to fidget and mew loudly.

Pulling out a box from behind the counter, he asked, "Excuse me, Mrs. S? Do you think the boys would like to color? We have some crayons and animal safety coloring books that I am sure they would enjoy."

Accepting his offer gratefully, the harried mother came up to the counter and handed each child a coloring book and some crayons. The boys quickly took them back to their seats and began coloring enthusiastically, but quietly. The receptionist also offered her the use of a light leash to clip onto the collar of the cat. "This will let her walk around without straying too far so you can relax and read a magazine while you wait for the doctor."

The client thanked him again and clipped the leash to her cat's collar. She was able to sit comfortably and read while the cat climbed up to sit on the window sill in the sun.

- How did the receptionist exhibit the traits for positive interpersonal relations and use them to diffuse the situation?
- How and when did he deal with his own prejudice?
- What might have happened if he had not intervened?

REVIEW

Match each of the six terms related to the communication process with its correct definition on the right.

1. message
2. sender
3. channel
4. receiver
5. feedback
6. reference points

a. form of communication such as writing or speaking

b. characteristics such as education and experience that determine how someone expresses and understands messages

c. the idea that one person wants to get across to another person

d. a return message

e. the person who first initiates a message

f. the person listening to a message

Select the best, most appropriate response for the staff on duty in reception to say in each situation.

7a. We all forget our checkbooks once in a while, sir, but payment is still due on the day services are rendered.

7b. Do not worry about forgetting your checkbook, Mr. Jones. You can drop off the check later or put it in the mail today. I can give you a self-addressed, stamped envelope if you like.

8a. I am sorry to hear about your kitten getting run over. You must be heartbroken. Cats are always running off to do their own thing. I am a dog person myself.

8b. I am sorry to hear about your kitten. It is heartbreaking to lose a pet.

9a. We would appreciate it if you would not smoke, Mr. L. We have oxygen in use and the smoke travels through the ventilation system to the patient wards.

9b. Of course we do not have an ashtray, Mr. L. Didn't you see the No Smoking sign?

Indicate whether the following statements are true or false. If false, what would be needed to make them correct:

10. Sitting with your arms crossed over your chest always means you are cold and trying to warm up.

11. To paraphrase, you repeat what someone said, word for word.

12. To provide the best possible care for the client and his or her pet, a staff member should attempt to determine the message behind nonverbal communication.

13. Body language always reinforces or agrees with the spoken message.

Paraphrase each of the following client's statements.

14. Client: My dog is sick. He will not eat, and he sleeps too much.

15. Client: My kitten has a cold. Her nose is running and her eyes are all watery.

Fill in the blanks with the correct answer.

16. The gestures and mannerisms a person uses to communicate are called _____.

17. When dealing with an angry client, you should try to keep a _____ attitude.

18. Preconceived ideas about a group of people made without taking individual differences into account are called _____.

Consider the following scenario: A client is shouting at the receptionist because he has been waiting 45 minutes to see the veterinarian. The receptionist is stunned by the man's anger, and shouts back, "Well, it is not my fault. Stop screaming at me."

19. What did the receptionist do wrong?

20. List six barriers to effective communication.

21. A co-worker mentions that you "preform" well under pressure. What is wrong with her pronunciation of the word in quotation marks?

ONLINE RESOURCES

Communicating in the Culturally Diverse Workplace (from Jobweb)

This article addresses the issue of cultural diversity in the workplace, including different styles of communication and how to implement effective channels of intercultural communication.

<http://www.jobweb.com>
Search Terms: Resources; Library; Workplace; Culture; Communications

Merriam-Webster Online Dictionary

Merriam-Webster offers this online version of its popular dictionary, including audio pronunciations to help visitors hear and understand how the word should sound.

http://www.webster.com

CHAPTER 6

Interacting with Clients

OBJECTIVES

When you complete this chapter, you should be able to:

- describe a professional telephone personality and demonstrate how to handle the types of calls received in an animal hospital
- describe how to use an appointment book and the computer to efficiently schedule a veterinarian's workday
- explain how interpersonal skills apply to client and patient reception
- explain what euthanasia is and when it is performed
- identify common client reservations and concerns about euthanasia
- describe client and staff reactions to pet loss
- identify stages and strategies for dealing with grief

KEY TERMS

screening
call director
station
call board
tickler file
wave scheduling
flow scheduling
fixed office hours
establishing the matrix
euthanasia

Introduction

Whatever your position is within the veterinary practice, when you communicate with a client, you are representing the entire veterinary team and it is important that you appear and behave professionally.

You will find yourself interacting with clients for a variety of reasons, including making appointments, greeting people as they enter, answering questions, and addressing other concerns. You should be seen by clients as a knowledgeable, helpful professional. When telephoning a client or answering a call from a client, you should know what to say, what not to say, and how to communicate in a friendly and effective manner. When speaking with clients in person you need to know how to address them, how to project your professionalism, and understand how your appearance affects your interaction. It is especially important to know what to do when discussing very serious matters such as a debilitating injury, illness, or euthanasia. Knowing how to interact with clients in a variety of situations is one of the most important parts of your career.

Telephone Management

One day, Mr. M's cat limped into the house on three legs. There was a considerable amount of blood coming from the injury. Mr. M did not hesitate to call his veterinarian's office to advise them he was coming straight in with an emergency. The telephone rang seven times before someone finally answered in a disinterested voice, "Hallo, can ya hold."

Before Mr. M could reply, he was put on hold. After a few minutes of wondering whether he had even reached the right number, he heard the same unidentified voice say, between obvious chews of food, "Can I help ya?"

"Is this the Animal Care Hospital?"

"Yup," replied the voice on the telephone.

"This is Mr. M. My cat is bleeding and it looks like her leg is broken. I called to let you know that I need to bring her in right now and I want to know if there is a doctor available."

"Why dinnit you say it was an emergency? I wouldn'ta put you on hold! Bring it in," the unidentified staff member said and hung up the telephone.

This, of course, is a lesson in how *not* to treat a client. If you were treated so rudely, you would probably take your pet-care business elsewhere.

The Telephone Personality

The telephone is often the first line of communication between the veterinarians and their clients, and it is where first impressions are made. As trite as it may sound, the reality is that first impressions are lasting impressions. Everyone who answers the telephone or makes outgoing calls must develop a professional and pleasant telephone personality. The employee in the example above broke every rule of telephone etiquette. As you read the following seven rules for developing a professional telephone personality, determine all the things the person who took the call did wrong.

1. *Answer promptly.* Make sure that the phone is covered at all times. Try to answer on the first or second ring and try not to let the phone ring

more than three times. There should always be another member of staff available to step up and take incoming calls. Every staff member should be aware of the telephone when it rings.

2. *Identify yourself.* Immediately give the name of the animal hospital or the veterinarian's name, for example, "Good morning, Dr. G's office. This is Kim; how may I help you?"

3. *Speak pleasantly.* Avoid extremes. Do not speak too softly or too loudly, too slowly or too rapidly. Pronounce your words distinctly and with expression; that is, do not sound mechanical, like a robot. Speak naturally, as you would when speaking to someone in person so that your friendly personality comes through in your voice.

4. *Be courteous.* In addition to being polite (please, thank-you, pardon me) give the caller your full attention. Never try to talk to a caller and someone else at the same time, whether it is the veterinarian or a client checking in for an appointment. Courtesy also calls for using plain language not the medical jargon you will learn while working in a veterinary office.

5. *Explain interruptions.* If you must leave the telephone to get information or put the caller on hold to answer another line, explain why and give the caller a chance to respond. Sometimes people *cannot* hold, and it is thoughtless to assume they can. Always thank a caller for waiting. Remember, a client's time is valuable too.

6. *Use the telephone correctly.* Place the receiver firmly against your ear and speak directly into the mouthpiece (Figure 6-1). Never eat, drink, chew gum, smoke, or put anything in your mouth, like a pen or pencil while you are talking on the telephone.

Photography by Kathy Nuttall, courtesy of Riverton Veterinary Hospital, Riverton, UT

FIGURE 6-1 It is important to hold the telephone correctly so that your voice can be clearly heard and understood as well as what the caller is saying.

7. *Conclude courteously.* Usually the caller has the option of ending the conversation. So, if you telephone, you would be the one to end the conversation. Always be polite and allow the caller to end the conversation first. You should always conclude your own telephone call with the same amount of courtesy.

BE PREPARED

Have an organized area for receiving and making calls. Include:

- Pens or sharpened pencils
- Memo pad or scrap paper for taking notes
- Telephone message pads
- Alphabetized, up-to-date telephone index of frequently called numbers
- A specific place for telephone messages (a spindle, a file, a tray, or the recipient's message page in the computer program

---

A truly pleasant and professional telephone personality will please you, your employer, and the client. Do not try to force yourself or it will not feel comfortable and your voice will sound stilted and rehearsed and the client may feel that you lack sincerity or are unsure of your responses. Treat people as though you value and appreciate them and it will become natural to you. Keep your main goals in mind: *to provide excellent service to clients and their pets and reflect the excellent standards of the practice.*

Review the earlier example and identify what was done incorrectly.

1. The person let the phone ring seven times, which is at least four times too many.
2. The veterinary office was not identified nor was a personal name offered leaving the caller to wonder if he had reached the wrong number.
3. When the call finally was answered, the caller heard an unpleasant, rude person speaking in a hurried voice as if the call was a major interruption. The grammar was lazy and inappropriate: "hallo" instead of "hello," "yup" instead of "yes," "ya" instead of "you," and "dinnit" instead of didn't.
4. The person answering the telephone was discourteous, telling the caller to hold without giving him a chance to respond.
5. The client was not thanked for waiting.
6. There was no explanation or apology as to why the caller was put on hold.
7. The call was ended very discourteously by implying that the caller was responsible for the delay.
8. The caller was told to bring "it" in instead of referring to the owner's cat.
9. The caller should have been the one allowed to conclude the call.

Screening Calls

The specific method of **screening** (handling) incoming calls will vary according to your employer's wishes. It is important to find out to what extent the veterinarian wants you to screen his or her telephone calls. For instance, one veterinarian might trust your judgment in deciding which matters need immediate attention and another might want all the details, minus any interpretation. Each type of situation requires a different approach. An appointment for a routine checkup is not handled the same way as a call about an emergency.

It is a good idea to work out a pattern for screening calls. A routine will help you obtain the information you need in a minimum amount of time while you maintain a friendly, warm, interested manner (Figure 6-2). By directing the conversation—politely—you can redirect any tendency the client might have to ramble and digress from the real purpose of the call.

First, you always need to know who is calling so that you can screen or "field" the caller if necessary. Most veterinarians accept calls from other veterinarians and immediate family members, no questions asked. In most offices, these calls will be transferred as soon as possible. If the caller does not volunteer a name, you should politely ask, "May I have your name, please?" or "May I ask who is calling, please?"

Next, you need to know why the person is calling so that you can take the appropriate action; screen the call, transfer it, take a message, or make an appointment. Whereas it would be impertinent to ask why Dr. T's husband wants to speak to her, it would be irresponsible *not* to ask why a complete stranger wants to. Usually a doctor prefers to accept only emergency or urgent calls from clients. Other calls will be returned when the doctor has time between patients or at the end of the workday.

Photograph by Kathy Nuttall, courtesy of Riverton Veterinary Hospital

FIGURE 6-2   While speaking with a client, maintain a warm, friendly, and interested manner.

Some clients call to ask the doctor for information or services that you could just as easily provide. If you do not immediately know why someone is calling, you can waste a good deal of time, as in the following conversation.

"Good afternoon, Dr. D's office, MJ speaking. How may I help you?"

"This is Tom G. Let me speak to the doctor, please."

"I'm sorry, the doctor is with a client right now."

"Well, I need to talk to her."

"Would you like to leave a message, so the doctor can return your call?"

"Just tell her I need an appointment tomorrow morning for my parrot."

If the veterinary assistant or receptionist had transferred the call to the veterinarian, it would have been transferred right back so that the caller could make an appointment. Some callers simply want to talk to the veterinarian and give no indication of their purpose. You can usually prompt these callers along by saying, "The veterinarian is with a client right now. May I help you?" If they still hesitate, you might ask if they would like to make an appointment, or would prefer to leave a message. The more specific you are in your questions, the more specific the caller will be in answering.

Another good reason to know why someone is calling is that you can prepare yourself for whatever action you need to take. For example, if you know that Mr. G wants an appointment, you can have your pen and appointment book ready or open the scheduling page on the computer program. If you know that a client is calling to find out when her dog was last vaccinated, you can quickly pull the file or check the patient's records in the computer. A well-organized member of staff can handle the average call in less than three minutes.

Now you know how to screen a veterinarian's calls, but what do you say when the veterinarian is busy, but not with a patient? Or what if the veterinarian is not in the office during regular office hours? Avoid the following statements, even if they are true.

- She is not in yet.
- He went home for the day.
- She went out to lunch.
- He is playing golf all afternoon.
- The doctor is busy.

Instead, simply say, "The doctor is not in at the moment" or "The doctor is unavailable." Then offer to take a message, and say, "May I ask the doctor to call you?" Remember, the veterinarian is entitled to a private life; you should never give clients a list of numbers to call in an effort to find the veterinarian. However, if the problem is urgent, you must try to reach the veterinarian or the designated emergency standby and assure the client of this fact.

Occasionally, you will answer the phone to find out that the caller dialed a wrong number. The caller may seem confused or insist you are the hardware store, even though you clearly identified yourself. To avoid a lengthy conversation, you should politely clarify things by saying something like, "I am sorry, but this is the Animal Care Clinic. Are you calling 555-5678?" This serves the double purpose of alerting the caller to the mistake and keeping her from repeating it. It is also worth remembering that this may be a client who has dialed the clinic by mistake or it could be a potential client. Always remain cheerful and helpful.

Making Appointments over the Phone

Much of your time on the phone will be spent making appointments and it becomes easier and quicker if you develop a pattern for asking questions. Surprisingly, a number of clients do not volunteer such basic information as their name, the reason for their call, or a convenient time for them to see the doctor. It is your job to *politely*, but efficiently, prompt such people along. For instance, after asking the client's name, if a client does not specify a time, you could ask one of these questions:

- Do you prefer morning or afternoon?
- Could you come tomorrow at three o'clock?
- When would you like to come in?

Of course, some clients present the opposite problem; "I *have* to see the veterinarian *today* at *ten*." If the schedule allows, you should accommodate the client's wishes. However, when the veterinarian is heavily booked already, you must make that clear while making the client feel wanted. "I am sorry, the only time Dr. T. has available today is 4 o'clock. Can you come in then?"

Sometimes the schedule is booked for a week or more in advance, and clients become upset because they cannot get a prompt appointment. In such cases, you might say that you will be happy to call them if there is a cancellation. In emergency situations, you should never offer to make an appointment, but advise the client to come in immediately regardless of what the appointment schedule looks like. Veterinarians and veterinary technicians should be alerted that an emergency is on the way in and relay to them the type of emergency the client has described. This helps everyone prepare. Many times clients who have an emergency do not take the time to telephone but rush into the clinic or hospital asking for help. Veterinary technicians and veterinary assistants should be immediately summoned to assist. As the team member responsible for the front desk, you should "stay put," do not leave the reception area open and unattended, but calmly gather together the all paperwork that will be needed. Many practices have a clearly stated policy regarding the order in which patients are seen. This is usually posted on the wall behind the receptions desk or in a sturdy frame on the reception counter. For example, *patients will be seen in this order: emergencies, appointments, and walk-in.*

Walk-in clients should never be treated as if they are a nuisance or made in any way to feel unwelcome. Explain to them politely that they will be seen as quickly as possible and if you can, suggest an approximate wait time. Never forget that they are there, acknowledge them often, and have them start completing client/patient information sheets. Be very aware that an emergency arriving unannounced is not the same thing as a walk-in client!

Taking Messages

Most offices have standard message forms for you to jot down the date, time, the name of the caller, return telephone number, and message (Figure 6-3). You should also include your own name or initials, especially in a large practice

To: *Dr. Morgan*

Date: *5/12*   Time: *12:30*   A.M. (P.M.)

WHILE YOU WERE OUT

Mrs. Thompson

of: *the Humane Society*

Phone:

555-0000

Need to confirm that Dr. Morgan will be their guest speaker.

To: *Dr. Brown*

Date: *5/12*   Time: *9:30*   (A.M.) P.M.

WHILE YOU WERE OUT

Mr. Richard Jackson

of: *client*

Phone:

555-1111

Regarding date and time for a kennel visit and now has a pup with possible abscess

FIGURE 6-3   It is important to obtain as much information as possible when taken a message.

where any of a number of people might answer the phone. Accuracy is important, and to minimize confusion, follow these guidelines:

- Write down every instruction clearly, to leave no room for misunderstanding.
- Read back, word for word, the entire message after you have written it down and confirm the telephone number where the caller can be reached.
- Establish clearly whether the message must be delivered immediately or acted on by a certain time or date.

Consider the "what-if" situations that could arise if the message is not acted on in time. This can be a difficult but important step. For example, suppose the veterinarian is out when a client calls describing signs or a condition that you recognize as dangerous for an animal. What if the veterinarian does not return soon? Should the client call another veterinarian who has been designated to take calls? Should you (not the client) try to track down the veterinarian? An important aspect of your job is to make judgment calls about how to handle "what-if" situations.

Coordinating Calls

Unless the office has a number of people answering phones, you are likely to face the problem of coordinating two or more calls at the same time. Most veterinary offices have more than one telephone line, and you may find yourself juggling calls while you are greeting incoming clients, accepting payment from others, and pulling or returning files. Handling several lines while performing other duties takes a bit of coordination—mental as well as physical. But you can handle the situation smoothly by mastering the use of your **call director**.

A call director is a telephone unit with **stations** that allow as many call to come in as there are open lines to receive them. The **call board** has buttons that represent the stations in your office. Station lights tell you how to respond to the call director. With some telephone systems, a slowly blinking light accompanied by a telephone ring alerts you to an incoming call; with others, the telephone simply rings. For almost all telephone systems, a constantly blinking light indicates a station (caller) is on hold while a steady light indicates a station (line) in use. When a button lights up and the telephone rings, push the station button down *before* you pick up the receiver and answer with your standard greeting. If you do not push the station button first, you could interrupt other calls in progress or the line will keep ringing.

What do you do when you are talking on one station and another call comes in on the call director? The procedure is really easy:

1. Politely ask the person you are talking with to hold. (Give the person a chance to respond in case she cannot hold.)
2. Push down the HOLD button, which is usually red.
3. Push down the station button that is blinking.
4. Answer politely, find out who is calling and why, and then ask the second caller to hold.
5. Push down the HOLD button again.
6. Go back to the first station, thank the person for holding, and finish the call as quickly as possible without sounding hurried or offending the first caller.
7. Return to the second station, thank the caller for holding, and proceed with the call.

To make certain your memory does not fail you, it is a good idea to jot down the name of the caller on hold.

Of course, this procedure can vary slightly, depending on the circumstances. For example, when the second caller is another doctor or your employer's spouse, you might transfer the call instead of placing the caller on hold. Each office has different policies for transferring calls. You will learn the standard procedure for your office during your training period.

Keep in mind that you might have to change your standard procedure if one of the calls you are juggling will take a long time to handle. For example, suppose that the second caller wants an appointment, and the first caller wants a summary of all the immunizations an allergy-ridden puppy has received in the last year. You will have to locate the first caller's file and spend several more minutes on the phone. In such a case, you may take the first caller's phone number and call back as soon as you have found the information.

Or you might ask the first caller to hold while you quickly schedule the second caller's appointment. If, however, you are away from any caller for a length of time, it is polite to return every so often with, "Mr. J, I am checking your records" or "I will be with you in a moment." Or ask if it would be convenient to call the client back with the information requested. The client may also have follow-up questions that need to be answered.

Sometimes more than one line rings at the same time. When that happens, follow the same procedure outlined for two calls. Simply ask each new caller to hold while you promptly complete the one in progress. You should generally handle calls in the order they come in but urgency or time considerations could justify a change in the order that they are handled. It is always better to finish one conversation at a time rather than switch back and forth between lines. This can be very annoying for the callers and difficult for you to keep things straight.

Sometimes people take a less personal view of veterinary care and, as with any other service, they shop around for what they perceive to be the "best deal." As above, you may quote fees for vaccinations and routine procedures. Callers who are simply price shopping may often say, "But Dr. X charges a lot less." This is a time to make your practice shine by explaining everything that is included in the fee, perhaps a free nail trim or health exam.

HOW TO HANDLE COMMON CALLS

If the caller wants ...	You should ...
information about a bill	retrieve the client's file, either the hard copy or the computer file and answer the questions yourself.
information about fees	follow office policy. In most veterinary offices, you can quote set fees, such as those for vaccinations and routine procedures such as a spay or neuter without complications. Usually, you can give a range for fees that may vary slightly but always explain the reason why there may be a variation in the fee. Make sure, in these instances that the amount is always given as an estimate and not a quote. Sometimes, you might have to explain that the veterinarian needs to examine the pet before determining a fee for other services.
to report on satisfactory progress or treatment	take a message and say, "Thank you very much for the update. I will make sure Dr. ___ gets this information."
to report on unsatisfactory progress or treatment	take a message and advise the client that either you or the veterinarian will return her call as soon as possible.
to request test results	with the veterinarian's permission, give favorable results, but also be prepared to answer questions and explain what each test reflects. If you are uncertain, transfer the call to a veterinary technician. Never try to bluff your way through. If the results of tests are unfavorable, the veterinarian will call the client so the patient's treatment plan and options can be further discussed.
to complain about care or fees	explain all the points you can, or offer to call back after you have checked the chart. If you cannot satisfy the client easily, it is best to have the veterinarian or the office manager return the call to the client.

Another version of the "price shopper" is a competing clinic or hospital. Handle these calls exactly as you would with any other caller: politely, cheerfully, and professionally.

Outgoing Calls

Answering the telephone and responding appropriately is only one part of acquiring excellent telephone skills. The other is the ability you also have to develop to place outgoing calls to clients or other business contacts. This task requires all the courtesy and tact of receiving telephone calls.

First, you should be organized. If there is more than one item that needs be discussed, you should write down a checklist to be sure each item is addressed.

Time is also a consideration, not only the time of day but also the amount of time you have available to speak without interruptions. The time of day is important for clients as well; be considerate and plan outgoing calls that are not likely to interrupt meal times or what may be a very hectic morning seeing children off to school. If you are placing a call to order supplies, it should never be done when the reception room is full and the telephone is ringing. Ask another member of staff to cover the front desk while you place the order on another extension where you are likely to not be disturbed.

Frequently, you will make outgoing calls to clients or another veterinarian your doctor wants to consult with. In this case, your role is straightforward. You could say something like, "This is MJ from Dr. D's office, returning Mr. Gs call." If the person is not expecting the call, you might say something like "Good morning, this is MJ calling for Dr. D. She would like to speak to you about Togo's latest test results." Then you will ask the person to hold for the doctor.

You might also make calls to:

- remind clients when boosters for vaccinations are due although these reminders are often sent to clients as reminder cards
- confirm appointments made far in advance
- advise clients that the veterinarian has an emergency, and ask them if they are able to come in a little later in the day or if they would prefer to reschedule their appointment.

Sometimes, you will be asked to relay messages from the veterinarian. For example, you might call to reassure a client that a pet is recovering well after surgery or an illness. Or if there is no office manager in the practice where you work, you might have the unpleasant assignment of calling clients to remind them to pay past due accounts.

Collection Calls

Most veterinary practices expect payment when services are rendered, but sometimes when credit is extended to a client it can become difficult to collect payments on the account. Some management consultants strongly advise against telephone collection calls because they may interrupt the client at an inopportune time or embarrass the client in front of others present. Sometimes, however, collection calls are unavoidable. Some clients may feel that promising to pay is all that is necessary. Unless you follow up, many clients still will not put the check in the mail.

You should *never* take it upon yourself to handle past due accounts without your employer's approval. Always advise your employer of any problems that develop. Never make a collection call within earshot of others, especially clients in the reception area. Not only is this in poor taste, but it could lead to a lawsuit against your employer. There are also criminal statutes and telephone company rules which you must be aware of and not violate. Violations of these statutes include the following:

- calls placed at odd hours
- repeated calls (considered harassment)
- calls to debtors' friends, relatives, employers, or children
- calls making threats
- calls falsely asserting that credit ratings will be hurt
- calls demanding payment for amounts not owed
- calls that frighten, abuse, torment, or harass another person

People frequently react in a hostile manner when reminded about a past due account, so try to approach the subject with sensitivity. Always speak directly to the person responsible for the bill. You should not leave a message with another person or on a telephone answering machine giving any details of the reason you are calling.

Telephone marketing and solicitation has become such a nuisance that it has prompted the enactment of the "no-call list." When people constantly receive unsolicited telephone calls, they may register their numbers on the national no-call list. Repeat offenders are subject to fines. Because of this, many telephone companies now offer a service that screens incoming calls to a private number. You may hear a recorded message that states something similar to this:

> This number does not accept unsolicited calls. If you are a solicitor, hang up now— if you are not a solicitor, please press 1

By following the voice instructions, your call will be put through normally.

Trying to collect past due accounts is not a pleasant task, but you can make it easier when you begin the conversation by identifying yourself in a clear and friendly manner with no suggestion of confrontation or aggressiveness. Avoid "small talk" asking questions such as "How are you doing?" and "How did you like the big blizzard last week?" The person you are calling will likely hang up on you, assuming it is the beginning of yet another telephone solicitor's spiel. The best approach is to be friendly and direct.

You may open the discussion by saying that you have been reviewing accounts outstanding and that their account has been inactive, that is, no payment has been received since. (Here, add the date of the last payment or when the client initially had the credit line extended.) You can usually avoid offending the person by assuming that the nonpayment is due to an oversight and say that you are calling as a reminder of the payment due and the amount outstanding. Your manner should remain pleasant and reflect your confidence that the client is not deliberately avoiding his or her obligation.

You: Your payment record with us has always been very good and we appreciate it. We were concerned, not having heard from you, and would like to discuss the status of your account.

Allow the client to respond, do not interrupt the client when an answer or explanation is being given. If the client pleads hardship or states a complaint, listen carefully, paraphrase the problem and when the client agrees with what has been said, promise that you will discuss the matter with your employer. Never show irritation in your voice or appear to be scolding the client. You simply want to find out why the bill has not been paid or there has been no response to the statements you may have previously sent. There may be extraordinary circumstances about which you would have no knowledge, such as a death in the immediate family or a debilitating illness. Never assume a client is just trying to avoid paying the bill.

The person will be more apt to cooperate if you indicate that the office can be flexible in a payment plan. For example, you can suggest a partial payment, followed by small monthly payments until the balance is paid. If the client promises to pay, ask when you can expect to receive the payment. End the conversation on a friendly note and be sure to thank the client for his or her time. Always follow up on the telephone conversation with a brief letter to the client confirming the details of payment that were discussed.

Documentation is very important when calling clients about outstanding accounts. You should always write down the date and time of the call and the date that the client has agreed to send a payment. The details of the conversation and verbal agreement should be entered into the client's file. In this way, you not only have a legal record (history) but you can more easily keep an eye on problem accounts. Most people are as good as their word, but occasionally follow-up calls may be necessary. Use a **tickler file** to follow up on the commitment dates. A tickler file is like an appointment book, only it has slots for mail and telephone messages that are organized by days of the month. Reminders are placed in the file on the day of the month concerned. Your computer software will also probably have an automated tickler program that will generate these reminders—to you—to follow up on accounts outstanding. Tickler files should be checked daily.

Remember that the goal of a collection call is to receive payments and clear outstanding accounts without losing or offending the client. An uncollected bill is insignificant compared to the damage that can result from an unhappy or offended client. Word of mouth spreads criticism faster than praise.

Off-Hours Phone Coverage

Someone must be available at all times to answer the telephone. Veterinarians usually subscribe to a professional answering service that provides coverage during off-hours such as nights, weekends, or holidays. The service screens client calls according to the veterinarian's orders. The last thing that should be done when the clinic closes for the night is to call the service and tell them that the clinic is closing and the calls will be transferred to them. One of the first tasks of the morning is to call the service and advise them that someone is now on duty in the clinic and will take the calls.

Instead of a service, some veterinarians use their own answering machine or a voice mail system in which clients can leave messages for the doctor or staff to handle directly. It may your responsibility to record the veterinarian's message on the answering machine.

Sample message: Hello. You have reached the Animal Care Hospital. We are sorry we cannot take your call right now. If you have a medical emergency, please call Dr. M. at 555-4321. If you would like to speak with Dr. T. or make an appointment, please call back tomorrow between the hours of 8:30 A.M. and 5:00 P.M. Thank you for calling the Animal Care Hospital.

Making Appointments

Scheduling Appointments

One of the most important duties of office management is scheduling appointments. You will be responsible for pacing the veterinarian's day according to his or her preferences. Schedule too many patients, and the result is chaos. Schedule too few, and the office loses profits. Your goal is to schedule the day so that no one's time is wasted, the clients are happy, the patients are well cared for, and the office prospers.

Types of Scheduling

To review, there are three general types of appointment scheduling:

1. *Wave scheduling.* With **wave scheduling,** the total number of patients to be seen in one segment is scheduled at the same time. For example, if the average time for each patient is 15 minutes, four patients are scheduled for each hour and are seen in the order of arrival. You might schedule four patients for the hour from 10 A.M. to 11 A.M. This method helps to adjust for any deviations from the schedule such as no-shows, late arrivals, or walk-ins. Although some patients may be seen later than scheduled, each hour is usually started and finished on time.

2. *Flow scheduling.* Using **flow scheduling**, patients are scheduled for 15-minute intervals. Provisions are made for longer appointments. This method usually results in the best control of scheduling and in a shorter waiting time for clients and patients. However, once you are behind schedule, it is almost impossible to catch up.

3. *Fixed office hours.* Clients come to announced **fixed office hours** with their pets whenever they wish. On arrival, they register or sign in and their pets are seen in the order in which they arrive. Obviously, with this system, it is difficult to control the flow of patients and make efficient use of resources and staff.

Whatever the type of scheduling, you always need to consider the time necessary for the examination or procedure and the availability of examination or treatment rooms.

Reminders

- Many clients like to receive a reminder when it is time for vaccination boosters. Reminders are usually sent about a month before a pet's vaccinations are due.
- Some offices use telephone reminders. A call saves postage; and even better, it guarantees that the client gets the message and may make the appointment at the same time. Many clients now prefer to have reminders sent via e-mail so it is a good idea to ask about their preferred method of contact. Client e-mail addresses should be entered into the client information file with their telephone numbers. Remember too, that not everyone has an e-mail address and some may wish to keep it private.
- Whether you remind by mail, e-mail, or telephone, always check the client's file first to make sure the pet has not been euthanized or that the owner has notified you of the pet's death.

The Appointment Book

Even though most hospitals and clinics now use computer scheduling (Figure 6-4), there are still many styles of appointment books available for use in animal hospitals and you should become familiar with both methods. These are designed with enough space for the number of appointments that can be

Photograph by Kathy Nuttall, courtesy of Riverton Veterinary Hospital, Riverton, UT

FIGURE 6-4 With the use of computers, appointments may be made from any station within the facility.

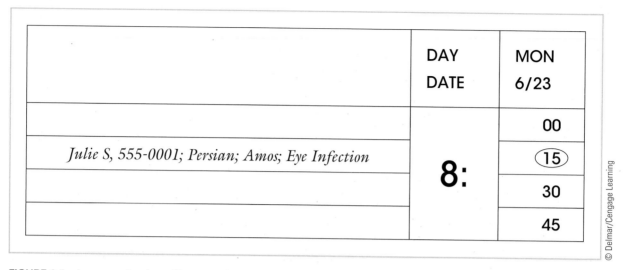

	DAY	MON
	DATE	6/23
		00
Julie S, 555-0001; Persian; Amos; Eye Infection	**8:**	(15)
		30
		45

© Delmar/Cengage Learning

FIGURE 6-5   An example of a written appointment book entry.

booked in a particular time frame (Figure 6-5). Each appointment slot has space for the following information:

- client's name and phone number
- pet's breed and name
- a brief notation for the reason of the visit

When setting up the appointment book, you should go through it immediately and block off the times when the veterinarian will not be available for appointments (Figure 6-6). Use red pen or pencil so that the marking stands out and appointments are not scheduled during times set aside for surgery or when the doctor will be out of the office. Unavailable times might include:

- holidays
- vacations
- personal appointments
- Sundays
- lunch breaks
- surgery
- days off
- evenings

This advanced preparation of the appointment book is sometimes called **establishing the matrix**.

The following suggestions should help you keep an accurate, effective appointment book:

- Confirm appointments taken over the telephone so that there is no mix-up. "We will see you at 9:30 A.M., Monday, May 9th."
- Confirm appointments the day before if they are longer than 15 minutes. This helps avoid lost time and revenue due to missed appointments.

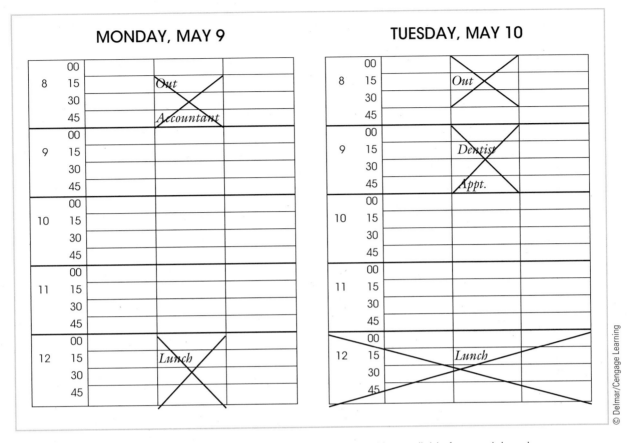

FIGURE 6-6 Remember to block off times when the veterinarian will not be available for appointments.

- Ask clients to please call if they are unable to keep the appointment or need to reschedule.
- Give new clients clear directions to the office so that they do not get lost and arrive late. If your office maintains its own Web page, advise them that there is a map and directions on the Web site.
- Be aware that walk-ins can become long-term clients and treat them accordingly, with the same courtesy as you do other clients. Emergencies are always given priority.
- Give appointment cards whenever possible (Figure 6-7).
- Keep the doctor aware of waiting clients and patients, especially when appointments are running behind schedule.
- Make sure you write every appointment down in the book, and always indicate when an appointment is canceled.
- Record no-shows and cancellations on the patients' charts and call or send notices to these clients asking them if they would like to reschedule.
- For cancellations, neatly draw a line through the original appointment information and give the time to another patient, writing the new appointment above the old. The appointment book can serve as a legal document if a client should sue the veterinarian, and erasing or whiting-out the canceled appointment could create legal complications. The same notations should be entered for electronic appointment scheduling.

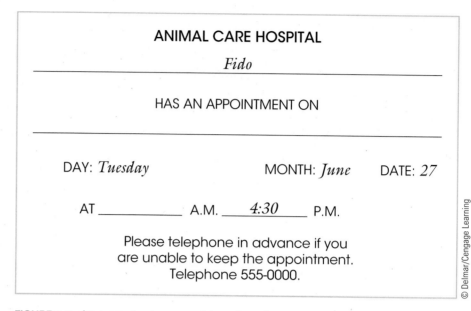

FIGURE 6-7   An example of an appointment card.

Organizing Appointments

The scheduling program in the computer is very similar to using the appointment book; the spaces for information and the data to be entered are the same. If the appointment scheduler, either book or computer file, has slots for four appointments every hour, should you schedule a patient every 15 minutes? Not necessarily. Appointments are scheduled according to the purpose of the visit, availability of examination and treatment rooms, and time the appointment will usually take. You should schedule each patient for the amount of time needed for that particular appointment. That time can range from just a few minutes for a vaccination to an hour or more if testing and radiographs are involved. Some lengthy procedures have periods in which the veterinarian's attention is not necessary, for example, while waiting for test results or a radiograph to be taken and developed. In those cases, you can schedule short appointments within the time frame of the longer one. You might also want to block off an occasional free period of 15 minutes to allow for any spillovers in the schedule. Sometimes clients request a specific veterinarian, and other times an appointment can be made for a veterinary technician for suture removal, for example.

To make the best possible use of the veterinarian's time, you should know how long each type of appointment usually takes and take it into consideration when you schedule the appointment. If you schedule carelessly—making appointments for patients without knowing the length of a time a particular type of appointment takes, you will probably end up with a full waiting room. This leads to clients who are dissatisfied, an unhappy employer, and increased levels of stress for all staff.

To make scheduling easier, take notes for yourself about the time it usually takes to complete different types of appointments. In this way, until you become familiar with the amount of time that needs to be allocated for specific types of appointments, you can tell at a glance if you can squeeze in an allergy shot for a cat in the afternoon, or if you will need to contact another client with a morning appointment to see if it can be rescheduled because an emergency has arrived. Veterinarians, like all of us, work at different paces, so you will have to check with the doctors regarding specific times that should be allocated for differing situations.

Scheduling House Calls

Clients usually bring their pets into the office for treatment. However, some clients may be unable to come in or to transport several dogs from a breeding kennel to the clinic. Many veterinarians have long-standing relationships with breeders and other facilities where house calls are more appropriate and time efficient. Another situation that can occur is when the owner prefers and requests that a beloved pet is euthanized at home. Scheduling a house call should always be made in consultation with the veterinarian concerned. Remember too, that in most instances, a veterinary technician will also be accompanying the veterinarian to the house call.

Most clinics or hospitals usually have a general policy in regard to home visits but you should always take a detailed message with the following information so that the veterinarian can schedule a time for the house call.

- full name of client
- name, breed or species of animal, and number of animals to be seen
- address, with clear directions to the owner's property
- phone number
- reason for the visit

As soon as you know when the veterinarian will be able to make the visit, call to let the client know. If, by chance, there should be more than one house call on the same day, the appointments should be arranged whenever possible to provide the most efficient route and travel time to the various locations. It is essential to get exact directions and telephone numbers from the clients.

Greeting Clients

In addition to making appointments and becoming proficient in the use of the telephone, another important aspect of the job is reception, welcoming clients, which is far more than sitting behind the desk and saying hello to clients. It is very important to greet and acknowledge every client who comes in regardless of the other tasks you may be dealing with at the time. It takes but a second to make eye contact and offer a quick, friendly smile in greeting. The client will know that you are very aware that he or she has arrived and will be taken care of as soon as possible.

Any number of events might change the pace of the day you so carefully scheduled. A seemingly routine ailment might turn out to be complicated and require more time than the scheduled, there may have to be a consultation with another veterinarian who has an emergency, or the veterinarian on duty might get held up in traffic. No matter what the cause, you need to learn to be flexible and to accommodate the unexpected. Regardless of the unexpected changes to the appointment schedule, you need to remain calm and cheerful to greet clients and patients and to soothe those in distress. In short, you are responsible for creating and maintaining a good social and professional atmosphere in the reception area of the veterinary office.

First Impressions

Would you buy groceries at a store with old, torn advertisements in dirty windows if you could go to a store with bright, new ads on sparkling clean windows? If you are like most people, you will assume that the old ads and dirty windows mean the groceries will be old and dirty too. Based on your first impression of the outside, you will pick the clean, neat store. The same applies to veterinary offices (Figure 6-8). A carelessly kept waiting room and reception desk suggests to clients that the veterinarian and staff are also careless (Figure 6-9). Who would want to trust their beloved pets to a seemingly uncaring veterinary practice? If the office is not kept clean, neat, and odor-free the practice will lose business.

The receptionist's duty starts before the first patient arrives and continues until after the last one leaves. You should arrive early enough to make certain the reception area is clean, neat, and odor-free, in addition to your other early morning duties. The first thing you should do is contact the answering service to let them know that someone is in the office to receive calls. Then you should have a good and careful look around the reception area; straighten up the magazines and make sure that all the refuse has been discarded and waste baskets are empty with fresh liner bags in place. Sweep, dust, and pick up things

Photograph by Kathy Nuttall, courtesy of Riverton Veterinary Hospital, Riverton, UT

FIGURE 6-8 The entrance to the practice should be clean and inviting.

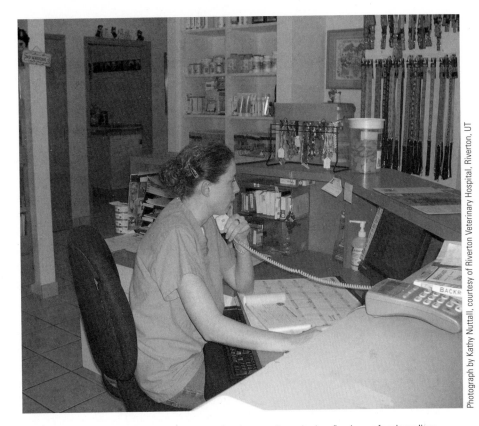

Photograph by Kathy Nuttall, courtesy of Riverton Veterinary Hospital, Riverton, UT

FIGURE 6-9   An uncluttered and organized reception desk reflects professionalism and allows client and patient needs to be readily met.

as needed. Tables and other furniture should be wiped down each morning with a disinfectant that helps to protect patients from any contagious diseases. The receptionist on duty should check the waiting area after every few patients and ensure that all is clean and tidy, exactly as it should be for the very first client. If this seems extreme, remember that every client—even the last one of the day—expects the office reception area to be clean and inviting. Even the most well-trained pets are likely to have "accidents" if they are scared and nervous. Everyone is responsible for cleanup, but if you are extremely busy you should call another member of staff to step in and help. Many clinics employ kennel staff to help clean up messes and tend to the cages, but you should never refuse to do it citing, "It's not my job." There are no circumstances which justify leaving urine, feces, blood, or vomit and not cleaning it up immediately. It may be unpleasant, but it should always be done cheerfully and without blaming the client or patient in any way.

A good start to the day can be with a thorough airing of the office. If there is no air conditioning, turn on ceiling fans or open *screened* windows (if you do not want to offer any opportunities for wild animals to enter or for pets to escape). This will help ventilate the area and remove stale air. The conservative use of deodorizers can help keep the air smelling fresh, but they should never be used to cover up odors. They should be air-neutralizers, not have strong scent of their own and be safe to use for pets, including birds. There are many specialized odor neutralizers available and you should only use the product chosen specifically for that purpose. Heat not only magnifies odors but makes clients and patients uncomfortable when the reception area is crowded. The waiting room thermostat should be set for no higher than 68°F or 20°C.

First impressions are not limited to the office itself and your appearance is just as important. When you greet clients, you should be just as pleasant and charming in person as you are on the telephone. No member of staff would *try* to irritate or offend clients, yet it happens. How do you ensure that you will make a favorable first impression when there is no way to tell what offends or irritates someone you do not know? There is no way to please everyone all the time but you can do your job best by focusing on two areas: professional appearance and good interpersonal skills. Always be neat and spotlessly clean, take pride in your personal hygiene, including clean hair and neatly trimmed nails. Do not use personal colognes or scented aftershaves as they can be just as cloying as an over-done air freshener and can cause many people to have adverse reactions. Keep a toothbrush at work and use it during the day rather than strong smelling breath fresheners. Do not chew gum or suck on candy!

Professional Appearance

After the initial phone contact, when a client first meets you he or she will notice how you appear, what you look like, and a bad first impression may be difficult to overcome no matter how skilled you are in performing your job. Most veterinarians ask their staff to wear some sort of uniform. It might be a scrub top worn over your street clothes, a combination of top and slacks, or a polo shirt with the clinic or hospital logo embroidered on it. Whatever your clothing requirements are, always make sure that your clothes are clean, neat, and in good condition.

Suppose you have taken a family member to a physician's office. How would you expect the staff to be dressed? Would you feel confident if you were greeted by a woman in a skimpy top and miniskirt? Would you be comfortable being treated by a doctor wearing jeans and a T-shirt? While these outfits might be fashionable at certain times and in certain social circles, they are never appropriate in a professional medical setting, including a veterinary practice. It is rare that a veterinary practice does not have a uniform or a dress code, but if there isn't one, use common sense when choosing your attire. It should be practical, easy to clean, and fit loosely enough for you to do your job. Casual does not mean sloppy. If uncertain, pay attention to the clothing other staff members wear and ask for their advice. You cannot refuse to do something just because it might ruin your clothes or get covered in pet hair. While jewelry can be attractive, loose fitting necklaces or bracelets can pose a danger to yourself and the patient; shiny dangling earrings might also prove too tempting a target for some patients, especially birds. You should not wear any jewelry which has the potential of getting caught up or harming the patient. A word about body piercings: those that are visible—for example, ear, nose, lip, and eyebrow—are frowned upon and do not present a professional image. They may not be permitted by your employer. Aside from the impression this type or ornamentation might give to clients, they present a real danger to you and to the patients.

Aside from your attire, people also will notice if you are a smoker. Smoking offends many people and it is a known health hazard. Veterinary offices ban smoking in or near the building because, aside from the other more personal concerns, oxygen is in use on the premises.

Think of every detail of your appearance and consider what kind of impression you make. Something as seemingly insignificant as your shoes could send an unfavorable message. For example, spike heels might suggest to a client that a woman is impractical and inefficient. A shirt unbuttoned to the waist might suggest that a man is unprofessional. Perhaps it is not fair, but if you want to succeed in your chosen field, you should consider what people think. *The care you take in presenting a clean, neat, efficient appearance will go a long way with clients who expect professional caregivers for their pets.*

Interpersonal Behavior

You might sound personable on the telephone and have a professional appearance, but still make a poor first impression on a client. For example, all you have to do is ignore someone who has just arrived. Put yourself in the client's situation and imagine how annoying it is to arrive on time, only to be completely ignored when you enter the office. Empathy—the ability to understand others' feelings—is an essential character trait for all veterinary staff.

The very *first* thing you should do when a client arrives is to acknowledge him or her. Even if you are on the phone or busy with something, acknowledge the client with a wave or a smile. If your duties require you to leave the reception area for a brief time when there is no one in the reception area, listen very closely for the door to open. There is usually a bell or a buzzer to announce that the office door has been opened. You should immediately return to the front desk to greet the person who has just entered. As soon as you return to the reception area, offer a personal greeting.

> RECEPTIONIST: Hello, Mr. J. I am sorry to keep you waiting. Would you like to sit down? Dr. T. will be with you in just a few minutes. Would you like to help yourself to some coffee?

For clients who have not been to the office in some time, verify their addresses, telephone numbers, and any other necessary information when they arrive and update all changes that need to be made on the client file and in the computer record.

For new clients, your greeting should be just as friendly. Begin by locating both the client's and the patient's name in the appointment schedule.

> RECEPTIONIST: Hello, are you Mrs. S and Rocky? It is nice to meet you. I'm MJ. I see Rocky is here for vaccinations. Let's just get some more information for your records and the doctor will see you shortly.

Of course, you do not have to use the same words each time. The important thing is to be natural, friendly, polite, and *genuine*. Reception is an opportunity to establish a positive relationship with both clients and their pets.

Being enthusiastic about your work will help make the veterinary office a more relaxed setting and more enjoyable for clients, patients, and other employees.

The receptionist needs to always be aware of clients and other visitors who are in the reception area and the reason for their visit. Often, sale representatives from pharmaceutical and supply companies will drop in on the off chance that the veterinarian is available. They should be greeted in the same friendly manner as any other visitor to the clinic, but under no circumstances should they be placed before a client. You may often hear "This will take just a minute of Dr. J's time" or "I just wanted to catch up with her about the latest development in pain control." Politely explain that the doctor is seeing patients and is unavailable and ask if they would like to leave their product information or a catalogue with a personal note for the veterinarian.

Clients and patients should always be seen in the order of their appointment times (or, in the case of patients who are a little late, their order of arrival) but there are exceptions. (Emergencies always take precedence.) Clients who are so late that the doctor really cannot fit them in can present a sticky situation. You do not want the client to get angry, but neither do you want to inconvenience other clients. If you really cannot fit the patient in, use tact in offering another time.

> RECEPTIONIST: I am sorry that your car broke down, Ms. B., Dr. T is in surgery now and we are fully booked for afternoon appointments. Could you possibly bring Felix in at nine tomorrow morning?

Clients whose pets are very ill or potentially contagious should be escorted to an examination room as soon as possible.

Handling Delays

Another important factor in a client's first impression of a veterinary office is the length of the wait. Waiting with animals can be uncomfortable for both the clients and their pets. It can also present a safety issue when a mix of strange dogs, cats, and other pets are together in a very confined space. When a client makes an appointment for 2:00 P.M., she expects to see the veterinarian at 2:00 P.M. An important part of a good relationship with clients is honesty about delays. Most people understand a 5- or 10-minute delay and will wait patiently. Any longer than that, and they watch the clock with increasing annoyance. You can avoid potential dissatisfaction by letting clients know when the wait will be considerable and why, an emergency for example. Offer the client the option of waiting, rescheduling, or perhaps leaving the pet for the veterinarian to see when he is available. There is usually no extra charge for this service, but additional paperwork will be required for "drop off" patients. Clients who will have the longest wait should be informed of the delay first. And, as you learned earlier, clients with later appointments appreciate your calling to inform them about delays of a half an hour or more.

Never be vague or noncommittal about how long a wait could be. If uncertain, check with the veterinarian so you will at least be able to give a reasonable estimate.

RECEPTIONIST: I'm sorry but Mr. J., Dr. T has an emergency surgery. She said it might be another 40 minutes to an hour.

If the client chooses to wait, you should offer whatever you can to make the wait more comfortable. For example, you might suggest the client take his pet for a stroll around the back of the building, provide a water dish for their pet, and offer the client refreshments. Clients appreciate a personal touch, which can make them feel recognized and valued not only as clients, but as individuals.

In addition to helping clients, your personal touch will make your own work easier, more pleasant, and more rewarding. For instance, if the office is always behind schedule, the front office staff should analyze the situation. Perhaps the scheduling is too tight and enough time is not being allocated to complete each appointment or maybe the veterinarian has trouble staying on schedule. Once the problem is identified, the staff can take steps to resolve it. Staff meetings present the best opportunity for brainstorming and problem solving.

Euthanasia

Euthanasia is a word derived from Greek. A literal translation is "good death" (*eu* means good and *anasia* refers to death). In veterinary medicine, the word means to end an animal's life humanely and painlessly. There are many reasons for euthanasia, for example, when an animal is suffering from a painful, incurable disease, has been severely injured, or perhaps when a beloved pet has become so old, weak, and frail that it is unable to eat and loses control over bodily functions. Euthanasia in these circumstances is most often considered a kindness, a gentle release from pain and suffering. However, there are many other reasons for euthanasia; the pet may be no longer wanted or the owners can no longer afford to care for it or the family situation has changed.

It is extremely important for all veterinary staff to understand the process of euthanasia, including those in office management as they are an important part of the team who talk to clients trying to face this difficult time. Staff must always approach these discussions with compassion and respect for the client and patient. You should never attempt to influence a client's decision or interject your own beliefs. The decision must be completely the client's own informed decision. With your help and ability to answer their questions, understand their fears and doubts, it will be the first step in accepting the reality of death for their pet. At no other time is your genuine care and empathy more important. The veterinarian, most likely, will have discussed euthanasia with the client, perhaps as a treatment option, but people—your clients—also need comfort from others, reassurance, understanding, and kindness.

There are various methods used to humanely perform euthanasia but the most common method for companion animals is by injection of a barbiturate drug into a vein. The barbiturate used is sodium pentobarbital. This drug acts rapidly by first inducing unconsciousness, followed by respiratory arrest (when breathing stops), and then cardiac arrest (when the heart stops beating). Death can occur within a matter of seconds. (*Note*: Sodium pentobarbital is a barbiturate and is a controlled substance. It should always be stored in the locked drug cabinet.)

Common Client Concerns

It is normal for clients to be frightened and confused, uncertain and hesitant when it comes to euthanasia, even if they have accepted that it is the best possible option for their animal companion. To make the situation a little easier, it is important to understand the concerns they have, and honestly discuss their concerns with them privately and without interruption. You should never have this discussion at the front desk. Always take the client to a separate room where you are both able to sit with each other. There should be no physical barriers between yourself and the client, such as an examination table or desk.

What Happens to the Body? A veterinary hospital usually has a contract with a licensed company that specializes in the disposal of animal bodies. Most often, the bodies are cremated as a group but the option for private cremation is usually available. In this case, the ashes of the pet will be returned to the veterinary hospital, presented in a small wooden box or an urn a client may choose. The ashes are held in the clinic or hospital for the client to collect. Reassure the client that the company whose service you use is of the highest ethical standard and great care will be given to ensure that the correct ashes are returned. Perhaps suggest to the owners that other clients have chosen a special plant or tree under which to place the ashes if they are to be returned. Some clients may wish to take the body with them for home burial and this should be gently discouraged. There are many regulations and laws that prohibit burying pets even on private property. (It is your responsibility to be aware of whatever laws apply in a given area.) There may also be a private pet cemetery in your area. As a part of excellent client care, you should already know if there is one, where it is located, the person to contact, and how the client can proceed in arranging a burial in a pet cemetery.

Being Present during Euthanasia. Clients may choose to be present during euthanasia. Some find it comforting to be there, both for themselves and the pet. Others feel the moment would be too much for them to bear. The option of being with their pet should be offered to the owner but the decision is made strictly by the client. If this is the case and the client wishes to be present, then you must explain what may also likely occur; sometimes there is what's known as an "excitement phase." This is a physiological reaction due to the drug and it does not indicate that the animal is afraid or is in pain but the pet may vocalize for a short second. At the time of death, it is also normal for the bladder and bowels to empty. When a client decides to be present, usually the patient is sedated first to allay even greater anxiety for the client or to reduce the potential of having difficulty restraining the pet for the injection. No one wants the last memory of a beloved pet to be one of struggling or to seemingly be "fighting back." Usually when a client is going to be present, the patient is taken to the treatment room where a catheter is placed in a vein in the rear leg. This access, or port, allows the owner to be close to the patient without interfering with the restraint and injection. Always cover the table with fresh, clean towels. Clients often bring a favorite blanket or bed for the comfort of their pet. If so, the clinic towels should line the bed and be gently removed from under the body and taken from the room. The owner should be given time to spend a few quiet moments alone with their pet after it has died.

Where the Euthanasia Will Take Place. Companion animals are usually euthanized in the veterinary hospital or clinic but some clients may request to have the euthanasia performed at home. Before agreeing to this, you must check with the veterinarian to determine when it can be done. It should be handled in the same manner as any other request for a house call appointment.

Scheduling The Appontment

Appointments for euthanasia should be made during the quietest times of the day or you may need to block as much as one hour in the appointment book. It could be during the last hour of opening or, if the practice closes for lunch, during that time. Grieving clients should never have to leave through a crowded waiting room. If a "quiet time" appointment just isn't possible, the owner and patient arriving for euthanasia should be promptly taken to a quiet and separate area. Many practices now have a small room set aside that does not resemble a clinical examination room but has been arranged with furniture and framed pictures that offer comfort. As difficult as it is in these situations, you still need to discuss the fee and method of payment and obtain a signed euthanasia consent form (Figure 6-10). The signed consent form should be

---

EUTHANASIA CONSENT FORM

VETERINARIAN _____

HOSPITAL NAME _____

ADDRESS _____

Client Name _____ Client Number _____

Pet's Name _____ Identification Number _____

Breed _____ Description _____

Owners/ Agent Driver License _____ Number _____ State _____

    I, the undersigned, confirm that I am the owner or authorized agent of the owner of the above described animal. I hereby unconditionally release the above animal for euthanasia.

Date: _____

OWNER OR AGENT SIGNATURE _____

Witness _____

© Delmar/Cengage Learning.

FIGURE 6-10 An example of a euthanasia consent form.

attached to the front of the patient file so that the veterinarian can see at a glance that the form has been signed and that this is the correct patient. The payment should be completed prior to the veterinarian meeting with the client. It is thoughtless to expect a client who has just euthanized a pet to have to stop and check out at the front desk.

Euthanasia Consent Form. Forms vary slightly depending on the practice but at the very least must all contain the above information. There may be an area for special instructions for the disposal of the body or the return of the ashes after cremation. Sometimes it may be difficult to determine who is a designated "agent" and can legally sign the consent form. It is a good idea when discussing an appointment for euthanasia to determine if the owner will bring in the pet or has made arrangements with another family member or friend to bring in the pet.

Sympathy cards should be mailed the same day. Many hospitals also arrange to have a single rose delivered to the home of the client. These extensions of care and sympathy should not be left longer than one day. Depending upon the relationship with the client, a telephone call in a day or two is also acceptable. You might say, "We have been thinking about you and wondering how you are doing." The call should express genuine concern, not one of a routine call-back. Always say "we are (thinking of you) concerned about how you are doing," not "I am concerned" or "I am thinking". Your telephone call should reflect the care and sympathy of the entire practice.

Grief

Clients normally experience pain, sadness, and grief at the loss of a pet. In many situations, the animal was regarded as a member of the family, and its loss is felt just as deeply. Being in such a powerful state of sorrow can also bring back other sad memories, making a person feel even more depressed. As a member of the veterinary team, you will likely feel a certain degree of grief yourself over the death of a patient, especially one you have been seeing for years, or one which has been in the hospital for care.

When someone loses a pet, whether through euthanasia, illness, or an accident, it produces strong emotions and different stages of grief which can be no different to losing a human family member or friend. It is normal and mentally healthy to work through the grief process.

- *Shock.* In many cases, the loss of a pet does not result from a prolonged illness, but can be due to a severe injury, short-term illness, accident, or a completely unexpected event. In these situations, people rarely have any time to prepare themselves for the loss of the pet, and may find themselves unable to accept the death of the pet immediately. They are in shock, trying to process the information and accept it as real.
- *Denial.* Shock can lead to denial, not truly believing that their pet and companion is dead. Some people feel the loss profoundly when they get home, and only then, when they are not greeted, do they accept that their pet is gone.
- *Bargaining.* Clients who are facing euthanasia may try and bargain with God or other spiritual beliefs in order to spare the life of the pet. It is a

psychological method of trying to keep hope alive. Clients might make promises to the pet or to themselves regarding the pet's care, food, or a promise of other special times together.

- *Anger.* It is a natural reaction for some people to become angry in the midst of a traumatic event. A client might be angry with the veterinarian and veterinary staff for not being able to save the pet, the person who hit the animal with a car or perhaps the person who left the gate open for the dog to run into the street or any number of people including himself.
- *Sorrow.* The clients you deal with will feel sad over the loss of their pets. This reaction is natural and normal; it is often accompanied by tears and you may shed a few of your own. While your instinctive reaction will be to try and comfort the person, be careful about touching clients. It is better to offer soothing words or condolences, and to listen if there is something the client wants to say. Be sure to keep a box of tissues available for grieving clients.
- *Depression.* While it is normal to feel sad after the loss of a pet, some people enter such a strong state of grief that they fall into a depression. Anyone still feeling an overwhelming sense of sadness and loss over an extended period of time should speak with a professional counselor. This can be very difficult to assess as everyone processes grief differently.
- *Guilt.* Sometimes people feel guilty, thinking that there might have been something they could have done to prevent the pet's death, or that the choice to euthanize was wrong. Left unresolved, feelings of guilt can prolong the grief recovery process.

You should be prepared to deal with a client expressing any or several of these emotions before, during, and after euthanasia. Try not to take any client comments or outbursts personally, as she or he may not be in an entirely rational state of mind. Be courteous, kind, and respectful. The most important thing you can do is listen and empathize.

Dealing with Grief. As with any other emotionally traumatic event, it takes time to get over the loss of a beloved animal. How long the recovery process takes differs from person to person. Helpful strategies that aid recovery include:

- Talking about the loss with a close friend or other family members.
- Talking with other people who have lost pets. Support groups exist in many areas, and your practice or hospital should have information about local resources. If not, do a little research and compile a list of local support groups, grief counselors, and online resources.
- Allowing themselves time to recover.
- Taking proper care of themselves, including eating right, and getting adequate amounts of exercise and rest. Illness can cause the grief period to last longer.
- Remembering the pet and what it meant to their lives.
- Getting another pet when the person feels ready, as a new pet can help a person to recover from the loss of a previous pet. The question of "when" varies from person to person, and in many cases, the answer is "never."

Professional grief counselors are available in many areas. The local Humane Society often has grief support groups. There are also a number of books about coping with the loss of a pet, as well as online support groups. The online resources at the end of this chapter may also prove helpful to you and to your clients.

SUMMARY

When you answer the telephone in a veterinary office, you become the voice of the practice to the person on the other end. You need to speak with the same respect and professionalism that you would expect to hear if you were the person making the call. You should be familiar with the office's method of screening calls, making appointments, and taking messages. When it comes to telephoning others, be sure you understand why you are calling, have all the information you might need to answer questions, and always be respectful, whether you are reminding a client about an appointment, asking a question, or requesting payment on a bill.

When making appointments, you should know whether your office uses wave scheduling, flow scheduling, or has fixed office hours, and utilize the appointment book and or the computer appointment scheduler to ensure that there is a record of exactly who is coming in when and why. Be aware of how long each type of appointment usually lasts, to avoid long waits for other patients. When there are delays, be sure the veterinarian knows patients are waiting, and let the patients know approximately how long a wait they can expect.

The reception area should be kept neat and clean, and you should always maintain a professional appearance.

Euthanasia is a medical procedure that requires all the staff to be professional, tactful, and compassionate. Special arrangements may need to be made when a patient is to be euthanized. Clients have strong emotional bonds with their pets and their reactions with the death of a pet can be very traumatic and painful for them. The office management team needs to know the range of emotions that the client might experience. Pet loss information and support group information should be made available to clients.

SCENARIO

The receptionist on duty at Dr. B's office answered the phone one morning. "Good morning, Dr. B's office. ST speaking, How may I help you?"

"Hello, this is Mr. M. I'd like to make an appointment for my iguana."

"Okay, how is Stilson doing?" (Important: the receptionist addresses the pet by name)

"Oh, he's fine, it is just time for his annual checkup."

"That's good to hear" replied the receptionist, while checking the appointment schedule for availabilities. "Do you prefer mornings or afternoons?"

"Afternoons are much better for me, thank you."

"Can you come in Thursday at 2:30?"

"That would be fine."

"Okay, I've got you down for Thursday, at 2:30" she replied, filling in his name in the appointment schedule. "Oh, our records show that you have an outstanding invoice from Stilson's emergency visit just over two months ago. Did you receive a copy of the bill?"

"Oh my, I forgot all about that! Can I bring a check with me when I come in on Thursday?"

"That would be fine, thank you. We look forward to seeing you again."

"Okay, see you Thursday."

"Goodbye, Mr. M."

- How did the receptionist's handling of the conversation help her to get the specific information she was looking for as quickly and completely as possible?
- Was it appropriate for her to bring up the outstanding bill while making the appointment?
- Which form of scheduling does this office most likely use?

REVIEW

1. Based on the following conversation that occurred at 11:45 A.M., fill out a message form like the one that follows the conversation. You may use today's date.

RECEPTIONIST: Dr. M's office. Anne speaking. How may I help you?

CALLER: This is M. Jones. I would like to speak with Dr. M. about my cat.

RECEPTIONIST: Oh, yes, Mrs. J. How is Tabby doing after her surgery?

CALLER: She has had a bit of bleeding around her stitches. I would like to talk to the doctor about it.

RECEPTIONIST: Dr. M is with another patient right now, but I will tell her as soon as she is free and have her call you. May I have your phone number, Mrs. J.?

CALLER: Yes, it is 555-1234. Please tell her Tabby bleeds whenever she moves.

RECEPTIONIST: When did the bleeding start, Mrs. J.?

CALLER: Eight this morning. When can I expect the doctor to call?

RECEPTIONIST: She will call you as soon as possible. For now, try to keep Tabby from jumping up on anything or going outside. If the bleeding gets worse, please call again immediately.

CALLER: Thank you. I'm really concerned.

RECEPTIONIST: Thank you for telling us this so promptly Mrs. J., if the bleeding continues, please, don't hesitate, just bring Tabby in. We don't want you to have to worry. OK?

CALLER: You are so kind. Thank you. I will watch her carefully and wait for Dr. M to call. Goodbye.

To: _____

Date: _____ Time: _____

A.M.
P.M.

WHILE YOU WERE OUT

Phone: _____

_____

Telephoned	☐ Please Call	☐
Came to See You	☐ Will Call Again	☐
Wants to See You	☐ Rush	☐
Ret'd Your Call	☐ Called Again	☐

Message _____

_____

_____

Signed _____

---

2. While the above two examples represent positive client/staff interaction, the person taking the telephone call in the following conversation makes mistakes in every response. Fill in what you think should be said. Make up any details you think are needed.

   STAFF: Hello, please hold.

   a. _____

   _____

   STAFF: (After three minutes) Yes, can I help you?

   b. _____

   _____

CALLER: I'd like to make an appointment.

STAFF: Just a minute. (Puts caller on hold for two more minutes.)

c. _____

_____

STAFF: Okay. What did you say your name was?

d. _____

_____

CALLER: I didn't. It's John Doe.

STAFF: How is next Friday at three?

e. _____

_____

CALLER: Actually, my cat got into a fight with another cat last night. He's got some deep puncture wounds I really think the vet should see him today.

STAFF: Oh, an alley cat on the prowl? Well, bring him in anytime, then, Mr. Doe. What did you say his name was?

f. _____

_____

CALLER: Tiger.

STAFF: Well, that figures! See you later. (Hangs up, laughing.)

g. _____

_____

3. The staff member in reception also makes mistakes in each of the following situations. Write what you think should be said or done.

   a. An incontinent elderly dog just had an accident on the waiting room floor. The person on duty in reception says to the client, "I am sorry, but could you please clean up the mess? The boy who usually does it is on break."

   _____

   _____

b. The receptionist is on the phone and pays no attention to the line of clients and pets waiting to check in.

_____

_____

c. The veterinarian has been unavoidably delayed for one hour, but the office staff chat with each other and say nothing to the waiting clients.

_____

_____

d. A client comes into the office to complain about a bill and yells at the receptionist who responds by getting angry and defensive and says, "If you cannot afford to keep your pets healthy, you should not be allowed to have them."

_____

_____

4. Create a log like the one that follows and enter the listed appointments from the doctor's schedule.

Office opens 8:00 A.M.

Surgical procedures in the early morning

Lunch from 1:00 P.M. to 2:00 P.M.

Personal appointment from 2:00 P.M. to 3:00 P.M.

Rounds to check on hospitalized animals from 4:00 P.M. to 5:00 P.M.

	00	_____
8	15	_____
	30	_____
	45	_____
	00	_____
9	15	_____
	30	_____
	45	_____
	00	_____
10	15	_____
	30	_____
	45	_____
	00	_____
11	15	_____
	30	_____
	45	_____

(*continued*)

(Continued)

	00	_____
12	15	_____
	30	_____
	45	_____
	00	_____
1	15	_____
	30	_____
	45	_____
	00	_____
2	15	_____
	30	_____
	45	_____
	00	_____
3	15	_____
	30	_____
	45	_____
	00	_____
4	15	_____
	30	_____
	45	_____
	00	_____
5	15	_____
	30	_____
	45	_____

Indicate whether the following statements are true or false. If false, indicate what is needed to make them true.

5. When a patient is euthanized, the client is required to take the remains.

6. Only medical personnel may be present during euthanasia.

7. Shock is a normal reaction to the loss of a pet from an accident.

8. A prolonged state of sorrow and grief might indicate a state of depression.

9. For most people, talking about the loss of a pet only makes grief worse.

10. Proper nutrition and exercise are important to the grief recovery process.

ONLINE RESOURCES

Mastering Telephone Etiquette (from Dummies.com)

This article from Dummies.com offers tips for proper phone usage when making and receiving calls, and when dealing with angry callers.

<http://www.dummies.com/>
Search Term: Telephone Etiquette

Pet Loss Support Pages

<http://www.petlossandgrief.com>
Association for Pet Loss and Bereavement (APLB)

<http://www.aplb.org>

CHAPTER 7

Stress

OBJECTIVES

When you complete this chapter, you should be able to:

- outline the causes of stress and the psychological defense mechanisms used to cope with stress
- list five positive ways to cope with stress
- explain how time management can reduce stress
- list five time management tools

KEY TERMS

stress

psychosomatic illness

National Institute for Occupational Safety and Health (NIOSH)

burnout

defense mechanisms

repression

displacement

projection

rationalization

intellectualization

sublimation

temporary withdrawal

malingering

denial

regression

prioritize

time tracking

time audit

Introduction

You work in a very busy animal hospital and suppose you have just been called in to work on your day off. Three of your co-workers are out sick with the flu, and two of the veterinarians are running a half an hour behind schedule. You already have two calls on hold, when you receive an emergency call that a client is rushing in with a dog that has been hit by a car. The dog is bleeding and the owner is hysterical. You are just about to instruct her on how to stop the bleeding when a German shepherd goes after another client's cat in the waiting room. The cat scratches its owner and gets loose. Two children are screaming and the whole of the office seems in chaos. Unlikely? Not really, every hospital and clinic will have days similar to this.

When you start your new job, you should expect to find yourself in a similar situation one day. There is no doubt that situations like this cause a great deal of stress and being able to cope with stress is not only an important part of a job but for your own wellness too.

What Is Stress?

Stress is the cause of the physical and psychological changes that occur in your body as a result of a change in your environment. Stress can be mental, emotional, or physical tension that results from frightening, exciting, challenging, confusing, endangering, or irritating circumstances—good or bad. Stress causes physiological changes such as a rise in blood pressure, a sudden rash, indigestion, stomach pain, constipation or diarrhea, frequent colds and flu, headaches, shortness of breath, or a nervous tic. It may cause emotional reactions such as crying, shouting, and becoming angry. Many diseases have stress as a root cause or an aggravating factor. Stress can lead to depression and **psychosomatic illnesses**: real physical symptoms resulting from an emotional and physiological response to stress. As a member of the office management team, you must learn to deal with stress from all directions: from your clients, your co-workers, your patients, and your own stress that comes from experiencing all of this. You may feel bombarded with everything and feel you just "cannot do this anymore."

Not all stress is bad for you. Many people perform job duties and other tasks at a higher level when they feel some pressure or stress. "Good" stress gives you the energy and desire to reach for goals and solve problems. A deadline can be a positive source of stress. Then again, if it is an impossible deadline, the stress can become overwhelming. The stress of reaching for a goal is positive when the mission is achievable, and you feel good when you have finished something. Failure to reach a goal can result in feelings of frustration, anger, and depression. It is important to have realistic expectations for yourself. Set goals and time frames that are realistic and achievable.

Causes of Stress on the Job

According to "Stress…At Work," a report issued by the **National Institute for Occupational Safety and Health (NIOSH),** the federal agency responsible for conducting research and making recommendations for the prevention of work-related illness and injury, the following job conditions can lead to stress:

- *Design of tasks*, including too much work, too few breaks, working too long, and too much emphasis on tasks that have little purpose and underutilize workers' skills and abilities.
- *Management style*, including supervisors ignoring the input of workers when making decisions, poor communication, and a lack of family-friendly policies.
- *Interpersonal relationships*, including a lack of cooperation and support among co-workers and managers.
- *Work roles*, including unrealistic or unclear job expectations, and giving an individual more responsibility than can be reasonably handled.
- *Career concerns*, including a lack of job security, little opportunity for professional advancement, and sudden changes without providing time for workers to properly prepare.
- *Environmental conditions*, including unpleasant or dangerous physical conditions.

Animal hospitals are not the easiest places to work. The demands of the job mean that you will have to interact with clients, co-workers with different personalities, and animals of all different types and behaviors. Some of the personality types of your co-workers will be difficult to deal with. Animals do not get sick on schedule—critical cases and emergencies always seem to happen at the most inopportune time, whenever you are behind schedule or the veterinarian is not available.

Your desire to work in a veterinary hospital or clinic is because you care about animals and their well-being. A major factor that contributes to your personal stress is in having to deal with the harsh reality that one is often faced with in veterinary medicine and these very traumatic situations sometimes can be very difficult. Some of the situations you will be faced with involve animal abuse and cruelty, animal abandonment and starvation, animals being rescued from a "crack house," and intentional drug exposure. It is not unusual, either, to arrive in the morning to open the clinic and find a carton at the front door with yet another litter of unwanted puppies or kittens. Dealing with situations like this can be difficult and cause a great of stress. Everyone in the practice wants to do the best for these animals, re-habilitate and find good homes for them but this is not always possible and oftentimes result in euthanasia, peacefully and humanely putting the animal down. Participating in euthanasia, the death of an animal, has its own level of stress. These issues are best handled with co-workers who care for and support each other, understand and empathize with each other, and learn to accept, as difficult as it can be, that this too is a part of veterinary medicine.

In order to cope with your job, you need to learn to be flexible, to prioritize and face the tasks one at a time not worrying too much about keeping everything under control at once. Realize that there are some things you can control and other things that you cannot. Recognizing the difference between the two is a very important step in learning to cope with stress.

Burnout

Burnout is a term used to describe a condition that results from too much stress over an extended period of time. In veterinary medicine, this can happen over months or years. Among all veterinary staff, burnout can result from feeling overwhelmed or becoming too involved with clients and their pets. For example, sometimes your care and concern is rewarded by a client complimenting the veterinarian instead of you personally and you take it as a personal criticism rather than praise for the entire practice and everyone involved in a patient's care. Or, you may totally over react emotionally to the death of a patient and the pain the owner is experiencing. When personally stressful events occur again and again a feeling of overwhelming anxiety develops, a feeling that you can no longer cope with it all, you become burned-out.

People who experience burnout generally feel a lack of control over their circumstances. They feel that no matter what they do, they cannot change things. The problem is chronic; that is, it will not go away. In attempting to cope, some people withdraw from interacting with others. They experience fear, anxiety, depression, with decreased energy and productivity and often feel ill or actually become ill because of the real physiological changes stress produces.

Coping With Stress

Coping Strategies

Work off stress with 20 minutes of physical exercise.
—From The Job Survival Instruction Book

There can be a very high turnover of staff in veterinary practices and a primary reason is burnout. You do not want to burn out, you want to grow and enjoy your new career and become successful. Fortunately, there are ways to avoid burnout with simple things like getting enough rest, eating healthy foods, exercising and learning to "turn-off," and having quality time for yourself. The following are some coping strategies that really work:

- *Exercise regularly.* Daily physical activity is a good release for stress.
- *Maintain good eating habits.* A healthy diet with balanced nutrition helps you to feel better physically, strengthens your immune system, and lessens your susceptibility to stress.
- *Deal directly with the source of the stress.* Do not let feelings build up inside of you, because this can actually amplify the effects of stress.
- *Slow down.* Try to limit mental and physical pressures. Do not become the source of your own stress.
- *Relax periodically.* Everyone needs to take a break, have a "time-out" even if it is only a quiet 15 minutes. We are all human, and we all become physically and mentally fatigued.

- *Join a support group.* Talking with other people who are experiencing difficulties with stress can be very therapeutic.
- *Discuss problems with a trusted friend.* "Keeping things bottled up" will only lead to further stress.
- *Volunteer.* Volunteering for even one day can be a very rewarding personal experience. Physical activities for outdoor projects, say a river cleanup or renewing a children's park, can be enormously satisfying and it does not have to be related to your work.
- *Learn to socialize.* Suggest social, casual times when everyone can meet for a pizza or invite another practice to join yours for a softball game and a picnic.

Some people use negative coping strategies as a quick fix for too much stress. Naturally, this usually makes things worse. Negative coping strategies include:

- *Calling in sick to "punish" the boss.* This only serves to punish your co-workers, the clients, and yourself.
- *Putting little effort into work.* Rather than reduce your stress level, this strategy is likely to increase it as your supervisor and co-workers will be dissatisfied with your performance, and you will generally end up making more work (and stress) for yourself.
- *Alcohol and/or drug use.* Alcohol and drug abuse can have very serious consequences in your professional, personal, and social life. It could also lead to dismissal from your job, become a danger to yourself and others and in some cases, to criminal prosecution.

Stress and burnout are serious problems that cannot always be avoided but by developing coping strategies, stress levels and the potential of burnout can be reduced and managed more effectively.

Defense Mechanisms

Psychologists and psychiatrists have identified a number of **defense mechanisms**, adjustments we make in our behavior, usually unconsciously, to help us deal with the experiences or feelings that cause psychological stress. Although you should never act the role of amateur psychiatrist, recognizing and understanding defense mechanisms can help you deal with clients and co-workers more easily.

- **Repression.** Socially unacceptable or painful desires or impulses are pushed out of the active and thinking area of the brain into the unconscious, actively unaware part of the brain without our even being aware of it. They are still there and may crop up in dreams or subtle behaviors. For instance, a woman may have forgotten her painful memories of being beaten as a child, but she winces with pain every time her husband clears his throat the way her abusive father used to.
- **Displacement.** Emotions about one person, idea, or situation are transferred to another, more acceptable or easier target. For example,

a man hates having to work for a woman supervisor, but he cannot do anything about it at work, so he is a tyrant with his wife and children at home.

- **Projection.** One's own ideas, feelings, or attitudes are attributed to someone else. For example, you convince yourself that someone else is to blame so you will not have to take responsibility and feel guilty. The man does not feel he is a tyrant with his wife and children; it is her fault that he has to crack down when he gets home because she is too weak and does not keep the kids under control.
- **Rationalization.** Actions are attributed to logical reasons without looking at the true motives of behavior. Suppose a man (or woman) has been trying to quit smoking, but after there is a blowup with the spouse and kids he or she says, "I haven't had a cigarette all day, so it is okay to have one now."
- **Intellectualization.** Again, reasoning is used to avoid the truth, as a way of denying strong feelings that may be socially unacceptable or difficult to accept. Instead of admitting to herself and to others that she is unhappily married, the woman chatters on and on about how well her husband is doing at work, how good he is at coaching Little League, and how much they are all looking forward to going on vacation.
- **Sublimation.** An instinctual desire or impulse is diverted into a socially acceptable activity. For example, a woman wishes she could see her old high school sweetheart, but instead she takes her kids to the beach a lot and reads romance novels.
- **Temporary withdrawal.** Ways are found to avoid dealing with a painful or difficult situation: a woman, for example, may watch a lot of television or another person may stop at a lot of bars.
- **Malingering.** Deliberately pretending to be sick enables someone to escape a situation that causes anxiety. For example, an employee sends word with a co-worker that he cannot go to his supervisor's dinner party because he has a stomachache, and then he orders a pizza with everything on it.
- **Denial.** To avoid accepting and dealing with a traumatic, stressful situation, the person refuses to admit or acknowledge that the situation exists. For example, the woman who was beaten as a child is in denial that she was abused, and the smoker is in denial that he or she is addicted to cigarettes.
- **Regression.** During times of high stress, we may return to an earlier mental or behavioral level. When someone is experiencing regression, the person returns to behaviors and seeks things, often food, which was enjoyed in an earlier, more pleasant time as a method of coping with depression.

You will probably recognize some of these defense mechanisms in yourself or in someone else. They are not usually a cause for great concern as these behaviors are a normal way of coping with stress. However, habitual use of defense mechanisms can indicate a need for counseling. Chronic dependence on defense mechanisms can point to interpersonal communication problems that might be solved if they were faced and analyzed.

KNOW YOURSELF

Before you can understand others, you must understand yourself. This quiz will help you take a good hard look at your interpersonal skills. Answer each question as it applies to you.

1. Do you find it easy to start a conversation?
2. Are you able to hold up your end of a conversation?
3. Do you ask good questions? Good questions are open-ended, requiring detailed answers instead of just yes or no.
4. Are you able to talk about things other than yourself?
5. Do you listen without interrupting the speaker?
6. Do you use body language when speaking?
7. Do you draw others into a conversation when they are reluctant to join in?
8. Do you avoid exaggerating facts when speaking to others?
9. Do you remember names of people when introduced?
10. Do you avoid using dialect, bad grammar, slang, clichés, or jargon in professional or formal situations?
11. Do you enjoy learning about people, their interests, hobbies, and ideas?
12. Do you keep others interested in what you are saying?
13. Do you give others an opportunity to express their views?
14. Are you able to discuss controversial matters without getting angry or upset?
15. Do you pay attention to the conversation without having your mind wander?

If you answered yes to at least 10 questions, your interpersonal skills are probably quite good, but try to work on any weak areas so that you can change no answers to yes.

PROBLEM SOLVING

Here are some problem-solving steps that you can use and pass on to others:

1. What is the problem? To identify the real problem, it helps to list the actual events or examples of behaviors that have contributed to the problem.
2. How can the problem be solved? Create a list of ways that the problem can be addressed. Books, counselors, and self-help groups are all good sources of information.
3. What would the outcome be if you did X, Y, or Z? To think through the possible outcome of various decisions, be logical and creative. It also helps to write things down.
4. Test alternatives. Begin with what seems to be the best course of action and continue down the list of options until a possible solution presents itself.

Time Management

One of the most common sources of stress is not having enough time in the day to do everything. Because it is impossible to add more time to the day, the best strategy is to find ways to save time and streamline work processes.

The following are some suggestions for improving and managing your time at work:

- *Reducing clutter.* In any practice, records, files, memos, and other papers can pile up quickly, making it difficult to find what you need when you need it. To fight clutter, keep yourself organized as you go. Return medical records and other files to their proper location immediately after they are no longer required. Set aside a separate tray for mail, memos, and other forms of correspondence. Whenever possible, use the computer to reduce the amount of paper used in the office (Figure 7-1).
- *Prioritizing.* Each day, make a list of what needs to be done, and then **prioritize** the tasks, tending to the most important items first and leaving the less important tasks for later in the day. If you run into difficulties or emergencies throughout the day, the items on your list that do not get done were the least important. This also helps you to avoid procrastination.
- *Time tracking.* When you are working on an involved project, it is easy to become engrossed in your work and not notice how much time has passed. To avoid spending more time than you planned on something, keep a small clock at your workspace and get into the habit of glancing at it periodically to note where you stand. **Time tracking** helps to keep you "on track" with the day's schedule.
- *Awareness.* It is important to be aware of what is happening in the practice or hospital. Is there something being worked on elsewhere that

FIGURE 7-1 By reducing clutter and learning to prioritize, you become more efficient and as a result, less stressed and overwhelmed than when things are allowed to pile up.

will be passed along to you to complete or work on? Is there something that is not being tended to that could grow into a problem? Being aware of what is happening in your work environment helps you to prepare for and prevent future timing difficulties.

- *Time audit.* If you try these strategies and they do not work, try keeping track of what you do each day and how much time you spend doing it. Throughout the day, take a moment to jot down on a notepad what you just did and how long it took you. At the end of the day, review your notes to see how you spent your time. What did you spend time doing that you did not need to do, or should have spent less time doing? This simple form of **time audit** can help you to identify areas in which improvement is needed.

SUMMARY

Working in a veterinary office can be stressful at times, especially when dealing with worried clients, the difference in staff personalities, and the concern over injured and hospitalized patients. It is important to recognize the signs of stress and develop effective coping strategies to help avoid burnout. It is also helpful to recognize defense mechanisms that help people deal with the effects of stress, both in yourself and others. They may be simple, temporary reactions, or possibly a sign of an interpersonal communication problem that needs to be addressed.

Time management can be an effective tool in reducing stress. Using methods such as reducing clutter, prioritizing, time tracking, awareness, and time audits can help you to make better use of your time.

SCENARIO

The receptionist was reviewing and purging old client files. She had just come back from lunch and was hoping to get through the remainder of the files by the end of the day. It was taking much longer than she had anticipated because the day was turning out to be very busy, with three emergency calls and two walk-ins, in addition to the regular workload.

One of the veterinary assistants came over and asked her for a specific client's file. The receptionist became very upset and said, "Why can't you ever get anything yourself? I don't have time to do your job, too!"

The technician replied calmly, "I'm sorry, but I saw that you were going through the files and I didn't want to mess up your system. Are you okay? I know today's been a little hectic. I could give you a hand if you're getting stressed out."

The receptionist replied sharply, "I'm not stressed out! I'm just trying to get this work done and people keep interrupting me!"

"I'm sorry. I'll get out of your way,"

"No, wait, I'm the one who's sorry. You're right, it's really busy today and I am feeling stressed. There's so much to do and there are so many people coming in, I just don't know what to do."

"Do all of these files really need to be purged today?" asked the technician

"No, not really, I was just hoping to get them all done, but it's really not that important. I'll do what I can today, and take care of the rest tomorrow."

"That's a good idea. Sometimes even two or three letters of the alphabet is all you can get through in a day. It is time consuming but not worth getting stressed over, OK?"

"Thanks. I just feel so much pressure sometimes, like I should be able to get through something and just can't."

- What conditions contributed to the receptionist's stress?
- What signs of stress do you think the technician noticed in her co-worker?
- What defense mechanisms did the receptionist employ to deal with her stress?
- How were coping strategies used to reduce her level of stress?
- How could time-management techniques have reduced the stress level?

REVIEW

Match each of the 10 terms with its correct definition on the right.

1. burnout
2. defense mechanism
3. displacement
4. malingering
5. projection
6. psychosomatic illness
7. rationalization
8. repression
9. stress
10. sublimation

a. pretending to be sick to escape a stressful situation
b. diverting instinctive impulses into socially acceptable activities
c. making logical excuses to avoid the truth
d. attributing one's own feelings to someone else
e. transferring emotion from the real target to another
f. pushing painful or unacceptable thoughts into the unconscious
g. unconscious behavioral adjustment
h. physical symptoms with a psychological root
i. bodily or mental tension
j. withdrawal caused by too much job-related pressure

Indicate whether the following statements are true or false. If false, what would be needed to make the statement true?

11. Not having enough time to do your job can cause stress.

12. Organizing as you go helps to increase clutter.

13. When prioritizing, put the least important tasks at the top of your list and do them first.

14. When doing a time audit, you should list each task you performed during the day by its level of importance.

ONLINE RESOURCES

The American Institute of Stress

The organization's official Web site, dedicated to advancing people's understanding of the role of stress in health and illness.

<http://www.stress.org>

Time Management (from Careers-Portal)

This article explains the importance of time management, suggests strategies for managing time better, and gives a short quiz to help determine if a person needs to manage time better.

<http://www.careers-portal.co.uk>
Search Term: Time management

Veterinary Technician Journal

The official publication of NAVTA, it has archived articles you can access on a variety of topics. One is "Stress, An Occupational Hazard."

http://www.veterinarytechnician.com

CHAPTER 8

Ethics

OBJECTIVES

When you complete this chapter, you should be able to:

- identify the rules of ethical conduct for veterinarians and members of the veterinary staff
- explain the types of behaviors and practices that are not allowed in the practice of veterinary medicine
- explain when the veterinarian–client–patient relationship begins, the responsibilities of those involved in it, and how it can be dissolved

KEY TERMS

ethics
jurisdiction
slander
confidentiality
substance abuse
veterinarian–client–patient relationship (VCPR)

Introduction

All members of the veterinary team, including the veterinarians and support staff must adhere to and promote the highest standards of ethical behavior. This chapter presents a general overview of **ethics**—the rules of moral conduct that all members of the veterinary staff should follow.

Putting The Needs of the Patient First

The veterinarian's job, first and foremost, is to relieve an animal's suffering, illness, or disability. The primary obligation of the staff is to aid the veterinarian in fulfilling this duty in the care and welfare of all animals. The AVMA (American Veterinary Medical Association) has developed standards of ethical behavior for meeting these obligations. In addition, each state provides its own guidelines which are defined in the Veterinary Practice Act. State veterinary medical licensing boards have the ability to review complaints of unethical behavior and have the authority to suspend a veterinarian's license to practice if the complaints are justified. In extreme cases, such as consistent complaints of animal abuse, the veterinarian's license may be revoked. State organizations for credentialed veterinary technicians have also developed standards of ethical behavior and have the same authority over complaints regarding veterinary technicians—that is, to suspend or revoke the credentials (license, certification, or registration) of a veterinary technician.

A behavior that is unethical doesn't necessarily mean it is illegal. Consider the following scenario:

A veterinarian has a long-standing relationship with a client who breeds, for example, Great Danes. It is the client's dream to produce a show champion and establish a breeding line. At nine weeks, the litter is brought in for ear-cropping as required by the American Kennel Club breed standard. The veterinarian performs the legal procedure. At nine months, the client returns with the "pick of the litter," a pup that in her estimation that has the greatest potential. There is one "small problem": the permanent upper incisor teeth are crooked. The client has noticed this with a littermate but as she doesn't intend to show the dog, it doesn't matter. During the examination, the veterinarian confirms that the alignment of the teeth is a problem but that the teeth might be realigned with orthodontic care, for example, the use of dental "braces," no different to those for children. The client is very enthusiastic about the procedure and the veterinarian agrees to consult with a friend who is an orthodontist. The client then makes an offer to the veterinarian: "If this works, I will make you a financial partner in the first litter this dog sires." The veterinarian readily agrees and is excited about the possibility of orthodontic treatment. In due course, with the orthodontist attending, braces are fitted to the dog's teeth. Some months later with other visits to readjust the braces, they are ready to be removed permanently. The dog now had perfectly aligned incisors and went on to win several championship titles. The first litter sired by this "champion" produced seven pups, three of which had the same genetic problem with the incisor teeth.

While the veterinarian did not break any laws in changing the appearance of the teeth, he clearly violated several ethical standards. First, by altering (disguising) a genetic problem that was passed on to a future generation and secondly, by agreeing to the partnership arrangement with the client. The client is also violating the ethics involved in breeding and promoting an animal that would not have met the judging standards and by offering a partnership to the veterinarian. It could be construed as collusion in unethical behavior.

Professional ethics is not governed by law but by doing what is right.

Right to Choose Clients and Patients

A veterinarian may choose to accept or decline any client or patient at any time, as long as the refusal of such service does not pose an immediate threat to the patient's life and does not result in prolonging the animal's suffering. However, once a person becomes a client the veterinarian, and by extension, the veterinary staff, must provide service. It is unethical and never acceptable to neglect the care of patients.

Emergency Care

Veterinarians, and only veterinarians, are allowed and required to provide essential services when an animal's life is at stake. Licensed veterinarians have a moral duty to provide care to save an animal's life or to relieve its suffering. Credentialed veterinary technicians may, with the direct orders of a veterinarian begin medical care in an emergency, such as placing a catheter for the delivery of fluids, applying pressure bandages to stop the bleeding, or to give an injection for pain or sedation with the exact dosage ordered by the veterinarian. The veterinarian must be in contact, either by physical presence in the facility or on the telephone with the veterinary technician. A veterinary technician should never begin medical treatment without direct orders from the veterinarian. When the veterinarian is not available, you should direct the client to the nearest veterinary practice or animal hospital where emergency treatment can be provided. Give the client exact directions to the other practice and provide the name and telephone number for the practice. You should keep the business cards of the nearest facilities readily available to give to the client and call the receiving clinic to alert them that an emergency will be arriving. If possible, tell them the type of emergency; you could say HBC (hit by car) or "sudden collapse" but you should never offer your own assessment or diagnosis of the condition to the client or to the receiving emergency hospital.

Obeying the Laws

It is illegal for a veterinarian or his or her staff to knowingly be in violation of the laws in the **jurisdiction** in which they practice veterinary medicine, to defraud clients or be deceitful in any way. No matter what your role is as a member of the veterinary staff, you should always be honest and fair in your professional relationships and know the laws which apply to the practice of veterinary medicine.

One example of obeying the law is with regard to wildlife. It is illegal for any member of the public (which here includes veterinarians and their staff) to harbor wildlife—orphaned, injured, or otherwise captive, that is, "found", unless that person is a licensed wildlife rehabilitator. Many times, wildlife is brought in to the veterinary facility by a member of the public who has somehow acquired or "rescued" an injured bird or mammal. By law, the veterinarian is only allowed to administer immediate life-saving care, to stabilize the bird or animal for movement to a licensed rehabilitation facility, or to perform euthanasia if, in his or her professional judgment, the animal or bird cannot be released back into the wild because of the nature of the injuries received. The exception to this, of course, would be if the veterinarian is also a licensed wildlife rehabilitator. Members of the veterinary staff are not permitted to take any wildlife home to feed and care for unless they are also licensed wildlife rehabilitators. The possession of different species is regulated by both state and federal laws. Every practice should have a list of wildlife rehabilitators in their area and maintain a professional relationship with them which includes further treatment and assistance if requested by the rehabilitator.

Misrepresentation

Veterinarians and veterinary staff should never professionally identify themselves as members of any organization by which they have not been licensed or certified. It is unethical for a veterinary technician, veterinary assistant, practice manager, or other members of the veterinary staff to misrepresent themselves and their qualifications. Veterinary technician, by definition, implies a certain level of education and credentialing by the examination board. If a veterinary assistant refers to himself or herself as a "tech," that is a misrepresentation of qualifications which could have serious legal complications.

Lending Credibility to the Illegal Practice of Veterinary Medicine

It is unethical for a veterinarian or any member of his or her staff to promote an illegal or unethical product or service offered by a nonprofessional organization, group, or individual. If you are aware of any illegal activities, including the practice of veterinary medicine by someone other than a licensed veterinarian, you should first discuss the situation with the veterinarian who should contact the appropriate authorities. For example, a friend of yours tells you about a person who makes house calls to vaccinate pets and treat small animals. This person is not a "vet" but has a great deal of experience from working in a pet store. He can get "shots and things" at a discount so it doesn't cost as much as the clinic where you work. Ethically and legally you are required to report this person who is practicing veterinary medicine without a license.

Responsibility to the Veterinary Profession

Duty to the Veterinary Image

Veterinarians and their staff members have a responsibility to enhance and uphold the positive image of practitioners of veterinary medicine in the eyes of colleagues, clients, and the general public. You should always act in an honest, courteous, compassionate, and professional manner, and remember that your appearance and behavior reflect directly upon the veterinarians you work for.

Duty to Professional Colleagues

It is both unethical and unprofessional to **slander** fellow veterinarians, regardless of your personal opinions. All members of the veterinary staff should refrain from making any negative remarks or implications about other veterinarians, veterinary practices, or veterinary hospitals. This also includes making any defamatory comments about any of the veterinarians for whom you work and other staff members. Conversations regarding any personal disputes with a veterinarian should be held in private, out of earshot of clients, co-workers, and the public.

Enhancing and Improving Veterinary Knowledge and Skills

Veterinarians, just as any other health professionals, are required to attend continuing education seminars in their field and should collaborate with their colleagues to share and expand their knowledge. Continuing education is also required for veterinary technicians, and is recommended for veterinary assistants. The more you know about animal care and your profession, the better able you will be to serve your patients and clients.

Duty to Society

Veterinarians and their staff benefit when they become involved in the community. There are many opportunities to contribute to public health efforts that relate to veterinary medicine, pet care, and animal welfare. The members of the veterinary staff should be available and willing to provide assistance as needed when public health is threatened.

Confidentiality

Every member of the veterinary team, regardless of the position held, has the responsibility of protecting client and patient information. The release of any client information, no matter how casual or innocent the conversation may seem is a violation of **confidentiality**. This breach of confidentiality could also be in violation of the law, result in a lawsuit, and the loss of your job.

Maintaining a Clean Mind, Body, and Spirit

If you suspect that a veterinarian or any other member of staff is under the influence of alcohol, drugs, or other substances that affect judgment, it is your ethical duty to report it to the most senior member of staff. While it can be a very difficult issue to deal with, it is a very serious issue and is cannot be taken lightly or ignored. You need to be 100 percent positive of your facts and be able to substantiate what you say. **Substance abuse** can lead to criminal charges. It is not a situation you should deal with on your own or take it upon yourself to advise the person to seek treatment. You should not discuss what your suspicions are with other staff members, but ask for a confidential meeting with another veterinarian or the practice manager who may also be aware or suspect that there is a problem. If you are uncertain about whom to approach and how, call a substance abuse hotline .Your call will be anonymous and confidential and these groups will be able to give you advice and guidelines on how best to approach the situation. Never make this sort of telephone call from the office, but from the privacy of your own home. You do not want this discussion to be overheard and create the potential for an even bigger problem. Anyone attempting to function with impaired judgment is particularly dangerous because the behavior and decisions that person may make could affect every member of the veterinary team, and the care of patients.

The Veterinarian–Client–Patient Relationship

The **veterinarian–client–patient relationship** (VCPR) is the professional relationship between the veterinarian, the client, and the patient. When this relationship is established, it is an understood agreement that the veterinary practice is there to provide appropriate, adequate veterinary health care, and that the client should follow the veterinarian's instructions for the patient. It should also be understood that the veterinary staff support and assist the veterinarian in providing care and to professionally serve the client and patient as directed by the veterinarian. As part of this agreement, the practice must keep and maintain medical records for each patient. The veterinary staff must properly update, file, and protect these records to facilitate optimal care and maintain confidentiality. The information must be kept confidential unless the owner of the patient consents to its release or a court orders the release of the information. While these medical records concern the patient, they are the property of the practice. However, said records should never be copied, distributed, or removed from the practice without the knowledge and written consent of the patient's owner.

Prescriptions

A VCPR is required for a veterinarian to prescribe medication. Veterinarians should never write a prescription for an animal that is not a patient, nor should any member of the veterinary staff dispense prescription medication, especially controlled substances, without a prescription from the veterinarian. Only a veterinarian may write out a prescription. It is important that requests for refills are approved by the prescribing doctor. If, for example, a prescribed antibiotic doesn't appear to be working, the patient should be reexamined by the veterinarian. Any request for a refill of a controlled substance should be carefully checked for the amount dispensed and the date the prescription was filled. Unfortunately, there is the potential of substance abuse by a client.

Termination of the VCPR

If there is no ongoing medical condition, the veterinary practice must notify the client that it no longer wishes to serve the patient and client and has decided to terminate the relationship. If there is an ongoing medical condition, the VCPR can only be terminated if the client is directed to another veterinarian for care. The veterinary practice must provide care until the transition is complete. Clients may terminate the relationship at any time for any reason and without notification.

SUMMARY

As medical practitioners, veterinarians must adhere to a strict code of ethics. So too must all the members of the veterinary staff. Each person should behave in a professional manner, and not allow any outside influences to affect the way the practice functions and offers veterinary care. The clients and patients of a veterinary practice are due the same respect and confidentiality that people are due in human health care.

SCENARIO

RJ was sitting in for the receptionist and was sorting the mail at the front desk when a woman burst through the front door, carrying a small dog that was bleeding badly.

"You have to help me, my dog is hurt!"

Rushing to the front door, RJ said, "Ma'am, please try to calm down and tell me what happened."

"I was walking my dog at the park, when this other big dog came along and they started fighting!" she replied. "We were able to get away, but he's hurt…we need to see a vet NOW!"

"She's bleeding pretty badly," RJ said, handing the woman a towel. "Use this to apply pressure on the wound. She's going to need stitches, and probably a tetanus shot."

"Okay, but I need to see a vet!"

"I'm sorry, but there are no doctors available right now. I'm writing down directions to the emergency veterinary clinic in the next town over, and I'll call ahead to let them know you're coming. They're not as good as our doctors, but they should be able to help you."

"Can't you do anything here? She might die if I can't get there in time!"

"Ma'am, there's nothing I can do"

"Do something!"

"Ma'am, I cannot help you. Your dog needs to see a veterinarian, and there isn't one here right now. I strongly recommend you continue applying pressure to that wound and take your dog to the emergency clinic."

"I can't believe you won't help me!" she replied, walking out the door. "Fine, I'll go somewhere where people care!"

- Which aspects of RJ's behavior were ethical?
- What should he not have done?
- What could he have done to handle the situation in a more ethical manner?

REVIEW

Indicate whether the following statements are true or false. If false, indicate what must be changed to make the statement true.

1. Veterinarians may only provide emergency treatment to their own patients.
2. Veterinary technicians should never identify themselves as veterinarians.
3. Veterinarians are morally and ethically required to treat each and every patient that requests medical care.
4. It is acceptable to question the ability of other veterinarians, but not the veterinarians that you work for.
5. Veterinarians, veterinary technicians, and veterinary assistants should all take ongoing education courses, attend appropriate seminars, and avail themselves of other educational opportunities.
6. It is acceptable to share patient information because veterinary patients are animals, not people.
7. A court order is always required for the release of patient records.
8. Veterinary team members should hold themselves to the same ethical standards as veterinarians.
9. Clients may end the VCPR at any time under any circumstances.
10. Veterinarians may end the VCPR at any time under any circumstances.

ONLINE RESOURCES

AVMA Principles of Veterinary Medical Ethics

This page contains the full text of the complete 1999 revision of the "Principles of Veterinary Medical Ethics" statement from the American Veterinary Medical Association.

http://www.avma.org
Search Term: Medical ethics

NAVTA Code of Ethics

Ethical standards for veterinary technicians, adopted by the National Association of Veterinary Technicians in America.

http://www.navta.net
Search Term: Code of ethics

PART 3

Financial Matters

CHAPTER 9

Fee Collection Procedures, Billing, and Payroll

OBJECTIVES

When you complete this chapter, you should be able to:

- explain how veterinarians' fees and fee profiles are derived
- recognize the different circumstances when credit should or should not be extended
- describe the rules and guidelines for extending credit on accounts receivable
- distinguish between legitimate and unadvisable reasons for adjusting or canceling a fee
- prepare billing statements and appropriately worded collection letters

KEY TERMS

encounter form	aging accounts receivable	Federal Wage-Bracket Withholding Table
travel sheet	telephone reminder	Federal Insurance Contribution Act
fee structure	written reminder	
fee profile	statute of limitations	payroll tax
cost estimate	Employment Eligibility Verification Form	social security number
credit bureau		Wage and Tax Statement
credit application	gross pay	
professional courtesy	salary	Employer's Quarterly Federal Tax Return
account history	net pay	
alpha search	exemptions	
cycle billing	Employee's Withholding Allowance Certificate	
accounts receivable ratio		

Introduction

One of the tasks in veterinary office management is requesting and accepting client payments. It is important to understand the way the billing process works, when fees should be adjusted or cancelled, and what to do when an account is past due. Another financial matter you should be aware of is payroll accounting, including what forms you need to fill out, what forms you should expect to receive, and when. Veterinary finances are an important area of practice management and can have far-reaching consequences for each member of the veterinary team if mishandled. As in any small business, the bottom line is profit. For a veterinary practice to thrive and grow, each staff member needs to be responsible and accurate in the financial management of the practice.

Fee Collection Procedures

While generally the job of the receptionist, collecting payments from the client may be handled by any member of the veterinary team who has received the proper training. Before the patient's file is taken to the examining room with the client, there should be a charge form attached to the file. This form lists all the procedures, vaccinations, and treatments which are provided by the practice. This form is sometimes also referred to as an **encounter form** or **travel sheet**; it "travels" with the file so that each treatment or procedure can be checked off as it is completed and no fees are lost. This form also serves as a statement of services for the client. The veterinarian marks the form to indicate the services rendered, and the client or veterinarian returns the form to the front desk to proceed with checkout and fee collection. If the veterinarian did not state different charges, then you enter the predetermined standard fees for the procedures that are checked. When the transaction is completed, give one copy of the statement to the customer as a receipt, and keep the other copy for accounting purposes. Almost all practices now use computer programs which allow the veterinarian to enter the billing charges directly into the client's file on the computer terminal in the exam room. This not only cuts down on the amount of paper and expense, but it is more accurate. On the hard copy or paper travel sheet, there is a greater likelihood that an item may be checked off in error or charges missed completely. The invoice is then sent (transmitted) from the exam room terminal to the front desk terminal where it is printed out for the client as a receipt. In this way, the patient's file is completely updated and there is no chance of misplacing the hard copy.

As with all your personal contacts with clients, you should be polite, friendly, and discreet when requesting and accepting payments. Usually, you will have no problem. However, some customers have other things on their mind after an appointment and forget to stop and pay the bill. You should not assume they are trying to skip payment. Nor should you let the bill go unpaid if your office requires payment at the time services are rendered. Politely call to the client, "Excuse me, Mr. P., could you come back for a moment?" Then, privately, say something like, "That will be $40 for today's visit, please," and present the statement. Never shout across the waiting room, "Hey, you forgot to pay!"

Understanding Veterinary Fees

Although you accept the money, the veterinarian sets the amount of payment for services rendered. Some veterinary offices follow a strict payment schedule of posted rates that are available to clients. Others have a range of fees that depend on the amount of time spent with the patient and the complexity of the diagnosis and treatment. The **fee structure** depends in large part on the types of services provided. For example, anesthesia for a surgical procedure might be billed by the minute or quarter hour, a vaccination might have a set fee, and treatment for an injury might include a combination of fee structures. When setting fees, the veterinarian must also consider the costs of maintaining the office and staff. Local veterinary associations and veterinary practice management firms can provide veterinarians with information about the fees similar practices charge in the same geographical area and across the country. This information is known as a **fee profile**.

In any case, you will have a fee schedule to go by, not only for completing the client's bill, but also in answering the questions that clients may ask about the costs. Clients, understandably, wish to know what their pets' medical treatment will cost and how the fees are determined. You should have a good understanding of how the fees are determined in the office—and good interpersonal skills—to discuss what can be an awkward subject. When complicated procedures are involved, fee estimates are usually provided by the veterinarian and discussed with the client beforehand.

Clients who may be wondering about the cost of treatment are not always assertive about asking questions. They may be reluctant to talk about money in any circumstance. When you speak with a client, particularly someone new to the office, ask a simple question such as, "Do you have any questions about any of our office policies?" This presents an opportunity for them to ask about fees without feeling embarrassed.

On occasion, clients may be unhappy with fees and do not want to talk to you about the issue or they may complain loudly when presented with the statement. The amount of the bill should never come as a surprise to the client. You should be patient with these clients and never argue or make demands. If there is a difficult situation, offer to have the veterinarian discuss the billing fees with them privately. The client may agree to this or realize that the amount charged is appropriate for services rendered. If the person persists, politely return the client to the exam room and have the attending veterinarian meet with the client.

The Costs Behind the Fees

It is important that all staff understand the high cost of setting up and maintaining a veterinary practice, which directly affect the amount of the bill. Operational costs often include:

- *Equipment and supplies.* Veterinary equipment—the laboratory equipment and diagnostic supplies, surgical table, lights and anesthesia machines, medical supplies such as bandages and suture material, and medications—is *expensive*. Radiograph equipment alone costs tens of thousands of dollars, and patient cages and isolation units can range from hundreds to thousands of dollars. Computers and software can easily cost a practice more than $10,000, and each piece of furniture or fixture in the practice can cost several hundred dollars. An average practice may spend several hundred dollars each month just on cleaning and disinfecting supplies and office maintenance services.

- *Mortgage or lease.* The space in which the practice is located must be paid for, in the form of lease or mortgage payments. If the property is leased, then the rent probably increases on an annual basis, and if the practice owns the building, property and water taxes must also be paid. These costs can run into tens of thousands of dollars per year.
- *Staff salaries.* The salaries of the employees working at a veterinary practice can range from $20,000 to $80,000 or more. There is also the cost of staff benefits such as health insurance, vacation pay, and uniforms. Veterinary fees cover the services provided not only by the veterinarian, but also those of technicians, assistants, receptionists, and kennel staff. Veterinary fees are the primary source of the practice's income.
- *Training and education.* The first thought that may come to mind when considering education costs is the veterinarian's cost of repaying student loans for his or her university and veterinary college tuition. Continuing education is required for veterinarians and veterinary technicians and this is an additional expense. There are also considerable costs related to training new employees in all aspects of their jobs. Clients come to a veterinary practice for medical expertise, but the knowledge and skills they receive take time and money to develop and maintain.
- *Miscellaneous.* There are numerous other costs such as professional journals and memberships, magazine subscriptions for the waiting room, advertising costs, telephone bills, professional liability insurance which increases the cost of coverage for the insurance on the property, the box of tissues on the receptionist's desk, the paper that is wasted and all the pens that disappear as if by magic.

When you consider the many costs incurred by a veterinary practice, the price of a single procedure pales in comparison. It is easier when you realize that the practice is not out to overcharge clients, but rather to cover its own costs in order to operate, make a profit, and continue providing veterinary care. Although you will not need to explain this to the client who is complaining about a bill, understanding the costs related to running a veterinary practice helps you to understand the cost of procedures and that, in turn, makes it easier for you to request payment when necessary.

Payment Planning

Any time services are rendered before payment is received, credit has been extended. The extension of credit is usually a very short time (minutes), because most clients pay their bills when they leave the office. But when the patient's treatment is ongoing, complicated, or especially expensive—such as surgery—credit is necessary to a greater extent, and a deposit for prolonged, expensive procedures is often required.

Clients have to pay all or a significant portion of these costs out of their own pocket unless they have purchased a pet insurance policy which covers the medical treatment needed. As part of the office management team, you can help ease the financial worries of these clients by preparing a **cost estimate** and with the approval of the veterinarian work out a payment plan that is reasonable for both parties. A common method is to make a fairly large initial payment to be followed by smaller monthly payments if approved credit is extended to the client.

Credit Bureaus. Before extending credit, especially for a significant amount, the client's credit rating (or payment history for other debts) should be verified. **Credit bureaus** collect information from various sources and provide information for a fee. Veterinarians can check the credit standing of their clients with other creditors. A bureau may contact the office about the status of a client's account. If a credit bureau contacts you, it usually means that they have permission from the client. This may not always be the case and you should verify this request with the client before releasing any information. If the client confirms and has given permission for a credit check you may provide certain information to the credit bureau: the date the account was opened, the current balance, and the highest balance at any time. You should not answer any personal questions about the client's character or paying habits nor should you volunteer any further information. It is often difficult to ascertain whether these calls are legitimate or not and it would best be handled by the practice manager who has the skills, education, and knowledge of the laws relating to the release of client financial information.

Credit Applications. Under the Equal Credit Opportunity Act, if the practice extends credit to one client, then it must extend credit on the same terms to other clients who request it. The only legal reason to deny credit is the inability to pay. You may be able to judge a person's ability to pay from the employment information on the client's registration form, which may also serve as an initial **credit application** in many offices. An employment history might be a good place to start, but circumstances change, sometimes rapidly, and the information on the client form may not be accurate. Any form used as a credit application may not ask the applicant's sex, race, color, religion, or natural origin. It may *not* ask for the applicant's age (but it may ask for the date of birth). The application may request marital status as "married," "unmarried," or "separated," but it may not use the terms "divorced," "single," or "widowed." You may be responsible for giving the credit application to the client, but it is not your responsibility or decision to determine if credit will be extended. Explain to the client that the application will be given to the office/practice manager and attach it, facedown, to the client's file. You should not stand at the front desk reading the application as it puts the client in a very uncomfortable and often embarrassing situation. Always be discrete; a credit application is not only confidential but contains personal and often sensitive information.

Truth in Lending. Under the Truth in Lending Act, enforced by the Federal Trade Commission, a veterinary practice, the same as any other business, must disclose the amount of finance charges to clients with whom you have an agreement to accept payment in four or more installments. The disclosure statement needs to be completed (Figure 9-1) even if there are no finance charges. The form must be signed by the client and kept on file in the veterinary office for two years. This law applies only when you bill the client each month for the amount of the payment due, not for the full amount owed. When the client is billed each month for the full amount owed and the client instead sends in partial payments each month, interest to the outstanding balance cannot be added after the fact. Every client who is extended credit, for any amount, should sign an interest disclosure statement.

Credit Cards. Most veterinary practices accept credit cards, not only to give customers more payment options, but also to avoid the sticky problem of extending credit and possibly having to collect delinquent accounts. Credit card

---

Sandra Smith, V.M.D.
100 Main Avenue
Anytown
Phone 555-5555

FEDERAL TRUTH IN LENDING STATEMENT
For professional services rendered

Customer _____

Address _____

_____

Patient _____

1. Cash Price (fee for service) $ _____
2. Cash Down Payment $ _____
3. Unpaid Balance of Cash Price $ _____
4. Amount Financed $ _____
5. Finance Charge $ _____
6. Finance Charge Expressed as Annual Percentage Rate $ _____
7. Total of Payments (4 plus 5) $ _____
8. Deferred Payment Price (1 plus 5) $ _____

"Total payment due" (7 above) is payable to _____ at above office address in _____ monthly installments of $ _____. The first installment is payable on _____ 20 _____ and each subsequent payment is due on the same day of each consecutive month until paid in full.

_____   _____
Date   Signature of Customer

©Delmar/Cengage Learning

FIGURE 9-1 Truth in Lending Statement.

WILL YOU TAKE A CHECK?

- A veterinary practice is a business. When accepting checks from clients the same procedures and policies used in any other business can guarantee that the check will be honoured and the payment received. Check verification machines are almost universal; in grocery stores, gas stations, retail stores, and, in short, any business that accepts checks in payment, including a veterinary practice. After checking all the details below, the check is "swiped" through the machine and the check will be verified electronically. If it is approved, a verification number will appear on the screen and you should write the number down on the reverse side of the check. Sometimes, the verification will be declined. There can be several reasons for this: perhaps the account has been closed, the checkbook has been reported stolen, or there are insufficient funds in the account. The check verification machine will not give any other information than the verification number (approved) or the word "Declined." If the check is declined, quietly advise the client that it has been declined and request another form of payment. It is a criminal offense to write "bad checks" and the majority of people would never intentionally do so. Never assume that it is deliberate and a way to avoid payment.

- Make sure the check is signed and completed correctly, with the day's date and amount and that the numerical and written amounts are the same. Do not accept post-dated checks.

- The office procedures policy manual will define the type of identification required prior to accepting a check, a driver's license for example. If so, compare the signatures and write the details of the "ID" on the reverse of the check. DL, for Driver's License, the number of the license, state, and expiration date. Accept out-of-state checks only with the approval of the office manager or veterinarian.

- Do not accept a third-party check, one that is made out to the client by someone unknown to you.

- Do not accept a check with corrections on it.

- Do not accept a check with "Payment in Full" written on it, unless of course the entire account really is paid in full. If you accept the check and the amount is less than the total balance of the account at the time, then the customer will not be legally liable to pay the rest of the balance. It is *illegal* for you to cross out the words "paid in full."

- As soon as you receive a check, stamp the back with the office's deposit endorsement so that no one else can cash the check if it is lost or stolen. Doing so will remind you to look on the back of the check for notations such as "Payment in Full."

- Do not accept postal money orders with more than one endorsement because then your endorsement will be the third, and two is the limit honoured by the postal service.

- Do not accept any check for more than the amount due. Some clients may ask you to take a check for more than the amount due so you will give them change. Do not do this for any client, even a friend who is well known to you. If the check is returned for insufficient funds, the office has lost the income not only for the billed amount but also the extra amount given out as cash.

company fees to the business may be as high as 8 percent of the total amount charged, but veterinary offices may not increase fees for customers who use credit cards.

If your office accepts credit card payments, there will be a credit card machine. During your training period your will learn how to process credit cards. The process involves the following steps:

1. *Clear the machine.* Check that the machine is ready to process a payment. To be sure, press the "clear" button.

2. *Enter the transaction amount.* When the machine is ready, punch in the amount of the charge (being sure that the decimal point is in the correct place—you may need to add two zeros to the end if it is an even dollar amount) and then press "enter." If you make a mistake, you can press the "clear" button and start again.

3. *Enter the credit card number.* Most machines have a "swiper" like those that you see at the register in many grocery stores and department stores. It allows you to physically swipe the card through the slot, letting the machine run the card automatically. If the machine will not read the card, you need to enter the information manually. You type in the credit card number and press "enter." You will then be asked for the card's expiration date, which is also keyed in. Press "enter" again and the machine will process the transaction.

4. *Approval.* If the card is approved, an approval number will appear on the screen. Most machines have a built-in printer that prints out the transaction automatically. If not, the credit card can be processed by hand with an imprint machine. The imprint machine uses standard forms in duplicate that are laid into the open face of the imprinter producing a carbon copy of the card on the transaction slip. What this method will not do is produce an approval number or decline the card, it is simply an imprint of the details of the card, the clinic account name and number, and an area for writing the details of the charge and the client's signature. Verify the client's signature and the expiration date of the card. As these forms are universal for most credit cards, you will also have to check on which type of card it is, for example, Mastercard, Visa, and so on. Finally, ensure that your office accepts the type of credit card presented. If using a manual imprinter, the office policy manual may state that you call the credit card company for verification.

5. *Signature.* Whether you use an electronic printout or imprint slip, the client signs the transaction slip on the appropriate line. After signing, the client receives the bottom copy and the top copy is placed in the cash drawer. You should keep credit card transaction slips separate from the checks and cash.

In addition to the credit card machine, some offices are now using computerized credit card programs that allow you to fill in the entries on the computer and then transmit them using a secure Internet connection. Regardless of the method for credit card transactions, you will receive training in the use of the machines and the office policy regarding credit card transactions. Failure to complete a transaction correctly can result in the practice not receiving payment, or if the decimal point is in the wrong place, the client may be over-charged or undercharged.

Adjusting or Canceling Fees

Fees may be adjusted or waived for a variety of legitimate reasons but they should not be adjusted as a way to avoid turning over an account to a collection agency. Fees should not be written off when the treatment has had poor results or if the patient died, a reason sometimes given by clients who ignore statements and refuse to pay. This may seem heartless at first, but consider the appearance of this sort of fee waiver or adjustment from the legal perspective. To some people, writing off the fee for this reason does not look like thoughtfulness and gener-osity; it looks like an admission of malpractice. However, the veterinarian may occasionally adjust the fee for a client who has disputed the fee if it is in the best interest of both to do so but only if the client is disputing the charges and not disputing the treatment. This offer should be made in writing, using the words "without prejudice" to give the veterinarian the right to reinstate the full charge if the customer does not pay the reduced fee within the specified time limit.

Owners may be genuinely impoverished or undergoing severe financial hardship. If so, a collection agency will be no more effective than you are in collecting the amount due. The veterinarian may decide to waive or reduce the fee in some cases. Payment must be discussed up front because it is always the veterinarian's decision whether to accept a patient when the owner cannot pay. Most veterinarians willingly provide service to a limited number of pet owners in these circumstances. However, a veterinary practice is a business, and no business can afford to provide unlimited free services to everyone who cannot afford to pay. If you find yourself in the situation of having to turn away a client, you should never turn them away without first helping them locate other veterinary care. You should maintain a list of the animal shelters and other organizations in your area that provide low cost or subsidized vet-erinary care for clients who just do not have the funds to pay. Recognize that their love and desire to care for their animal companion is no less than those clients who may be better off financially.

The client and the veterinarian should have an agreement in writing detail-ing the terms of the reduced fee before services are rendered. When the client undergoes unexpected financial hardship in the middle of an animal's treat-ment, the veterinarian may also choose to reduce the fee. In this case, the full amount should be entered on the billing charge and when the client has paid the agreed-on amount, the rest can be written off as an adjustment. This invoicing method should be explained to client and included in the written agreement to avoid any misunderstanding.

Another reason for fee adjustment is **professional courtesy**. Veterinarians frequently offer their services to other veterinarian without charge. For exam-ple, a veterinarian may need a consultation with a specialist or to call in another veterinarian who is very experienced in fracture repair, or perhaps a visiting veterinarian owns an exotic pet or bird and takes his animal to the exotic animal veterinarian for care and treatment. Many veterinarians provide medical care and medications for their employees' pets either without charge or at a discounted rate as a part of the employee benefit package.

Billing Clients

Most veterinary practices have a policy that states that payment is due at the time services are rendered. For routine office visits and relatively inexpensive procedures, it is reasonable to ask for payment at the time of service. Occasionally it is sometimes necessary to extend credit, for example, when the treatment is prolonged and intensive or the surgical costs are high. When credit is extended for a client you may be responsible for sending monthly statements. For some clients, the invoice is reminder enough and they pay on the bill regularly and promptly. Other clients may not be so conscientious and when at the end of 30 days no payment has been received, a statement should be sent to the client. Statements also need to be sent every 30 days to clients whose account has a debit or credit balance of one dollar or more. If a client has a credit balance, that is, money which has been overpaid to the account, the money is usually applied to the client's next visit. If there is a small amount outstanding, say less than $10, it makes very little financial sense to continue to send statements when the cost of postage soon meets the amount of money outstanding. For accounting purposes, small amounts are usually written off as a loss.

Every veterinary office has established guidelines for bill collection, including when it is acceptable to extend credit, when payments are due, how the client is billed, what procedures are followed in collecting delinquent accounts, and how long a delinquent account will be carried before it is turned over to a collection agency.

Thanks to the proliferation of computers and easy-to-use accounting software, account records are easier to maintain and statements are generated directly from the client files. The entire **account history** should have all the same information that you would have on a hard copy for each client: name, address, pet's name, and so on.

The benefit of computer accounting is its versatility. Software has been designed to perform almost every financial function. If all charges and payments are entered correctly in the computerized file a current balance for each client is available at the touch of a button. While the computer contains a client's complete financial records as well as his or her pet's clinical records, the program in the software will allow you to print only the information you need for billing purposes just as it produces the original invoice for a client's visit.

There are several ways to locate a client's file. Veterinary software programs provide access to files either by entering the account number or the client's name. When you know the name of the client the easiest method is an **alpha search**, or typing in the first few letters of the client's last name.

On the monitor will be a list of names beginning with those letters, and you select the right name and the program will pull up the correct file. From this, you can make whatever changes are needed and print the statement directly from the file. Computer programs can be set up to age accounts and sort them so that statements can be prepared for only those accounts you wish to bill. These programs also itemize the portions of the account that are current, 30 days past due, over 60 days past due, and so on.

One of the reasons a computer can work so fast is that it can be programmed to accept brief codes. For instance, the computer software may be programmed to recognize abbreviations. Once programmed, the computer recognizes the code and automatically produces the entire word. Brief codes can also be programmed to indicate the method of payment: for example cash, C, check, CK, credit card, CC. The codes used by a specific practice may differ slightly from those used in another type of practice but all use standard medical terminology abbreviations.

Computers also eliminate the possibility of making a mathematical error, providing of course, that there are no typing errors. You should proofread any entry you make before saving and updating the file. Mistakes found later can still be corrected, but the original entry is retained as a safeguard against potential embezzlement—stealing money from an employer through fraudulent accounting practices. For example, no one can enter a transaction, then delete it and keep the money.

Sending Statements

Clients are more likely to pay promptly if they receive statements at the same time each month as most people habitually pay all their incoming bills at the same time. "Bill paying day" usually coincides with their own paychecks. Also, the Fair Credit Billing Act states that the date of mailing statements for credit accounts cannot vary by more than five days unless the debtor has been notified about the change. Since many people pay their bills at the beginning of the month, it is a good idea to mail bills by the 25th of every month so that they are received at the appropriate time. If there many statements to send, there may not be time enough to prepare them all on the same day. It is still possible to maintain the same cycle each month with **cycle billing**. This is a method that is often used where accounts are grouped alphabetically and divided into the number of statements that need to be sent. For instance, if you set aside two days a month for billing, you could send statements to clients with names beginning with A through M on the 10th, and N through Z on the 25th. Or, if billing three days a month, you could mail A through F around the 10th of the month, G through M around the 20th, and N through Z the last day of the month.

Collecting Past Due Accounts

A small percentage of clients fail to pay their accounts on time. Some are simply negligent; they lose track of time and let their bills go until they get a reminder or two that the account is delinquent. Some are unable to pay because of financial difficulties and a few others are just not willing to pay. A veterinary office is doing pretty well if its **accounts receivable ratio** is less than two months. That means that the total of all accounts receivable should equal no more than two months' worth of gross charges. (However, in offices that extend credit extensively, a ratio of up to three months may be acceptable.) When the ratio gets higher than desirable, it is time to become more aggressive with account collection procedures.

The accounts receivable ratio can be calculated by using the following formula:

$$\text{Accounts Receivable Ratio} = \frac{\text{Current Accounts Receivable Balance}}{\text{Average Gross Monthly Charges}}$$

Example:

Annual Gross Charges	$144,000
Average Monthly Charges	12,000
($144,000 ÷ 12 months)	
Current Accounts Receivable Balance	$18,000
Accounts Receivable Ratio	1.5 months
($18,000 ÷ $12,000)	

In this example, the accounts receivable ratio is one and a half months. The average monthly charges are $12,000, and the current amount outstanding is $18,000. If the current accounts receivable were $24,000, which is the equivalent of two months of monthly charges, the accounts receivable ratio would be two months ($24,000 ÷ $12,000 = 2).

It is important to try to collect on all accounts receivable for several reasons. The veterinary practice is a business that must bring in income to pay expenses and continue to provide services. Delinquent accounts also negatively affect the clients who owe the money, because they may be embarrassed to bring their pets into the office for medical attention when they really need it.

Aging Accounts Receivable. Analyzing accounts that are past due is called **aging accounts receivable**. Aging begins on the first day of billing, which depends on the billing cycle. It may be the day service was rendered or the day of the first mailed statement. Always keep track of all accounts receivable so that you can use the appropriate collection methods which should be defined in the office policy manual. Bills that have been outstanding for less than 30 days are considered current and require no additional attempts at collection. A bill that has been outstanding for 60 days is past due and merits a *tactful* and *friendly* reminder.

Telephone Reminders. A telephone call, if made at the right time and in the right manner, is often the most effective method of debt collection. (See the section on collection calls in Chapter 7.)

Written Reminders. A reminder can also be put in with the statement either in the form of a sticker or note attached to the statement, or a printed-out form letter that is personalized for the specific account. Whatever you do, *do not* invade the client's privacy by using a postcard to communicate your message, and *do not* mark the outside of the envelope with a notation such as "overdue notice." When writing a reminder or collection letter, avoid using phrases such as "You *forgot/neglected/failed* to pay your bill" or condescending language such as "I am sure you just *overlooked* this bill." Each practice will have its own preferred reminder letters to send.

When a bill is 90 days past due, you should give a second reminder. Your computer program can probably be set to alert you when invoices are 60, 90, or 90+ days overdue. Most clients, after receiving one or two friendly reminders, will contact you to explain why they were not able to pay sooner or to set up a mutually agreeable payment plan. Once an account is past due 90+ days, it has become a collection problem and more persuasive collection techniques must be used.

Forced Collection

It is important to follow up on early friendly reminders with more forceful, though still tactful, reminders. Most offices use a series of three, four, or five letters and/or phone calls that become increasingly firm in their request for payment. After 90 days, it may be office policy to send a "final notice" indicating that unless the account is paid by a specified date, it will be turned over to a collection agency or, less often, that a lawsuit will be filed in small claims court. Even if you have used the telephone for prior communications, the final notice must be in writing and it should be sent by certified mail so that the recipient has to sign for it. The signature proves that it was sent and delivered. You should never send this notice unless the veterinarian has instructed you to do so. It should not be an idle threat; the practice owner/veterinarian must be prepared to follow through on any warning of further action that will be taken. Idle threats may be considered harassment and the client may be able to sue.

Turning an account over to a collection agency is not a pleasant task, but it is usually the wisest solution providing that the client has had every reasonable opportunity to pay the bill or arrange for payment. Delinquent accounts should be turned over to professional bill collectors on the date stated in the final notice. Consider the law of diminishing returns: Beyond a certain point, the odds of recovering the debt drop lower and lower, and the time that is spent in debt collection will not be worth whatever funds, if any, that the practice receives. A reputable collection agency may be able to collect on a higher percentage of these accounts, but if the client genuinely has no money, the collection agencies will have no better success than the efforts of the veterinary practice. If payment is successful through a collection agency, the fee that is charged for their services is usually 20 percent to 50 percent commission on the amount collected.

When an account is considered a 100 percent loss and all other methods have failed, yet the client can clearly afford to pay, another option the veterinarian may consider is to take legal action. However, litigation is so costly and time-consuming that most veterinarians are more likely to write off the account as a bad debt than to take the client to court. Any legal recourse must be exercised within the **statute of limitations**, the time limitation dictated by state law.

Finally, this point is so important that it bears repeating: Your employer has specific policies regarding bill collection. *Always follow office policy and protocol for billing and collections.*

> ### USING A COLLECTION AGENCY
> - When the veterinarian decides to turn an account over to a collection agency, the agency will need the following information:
> - Debtor's full name
> - Name of debtor's spouse or any other person who might be responsible for the bill
> - Last known address
> - Full amount of debt
> - Date of last credit or debit on the client's account.
> - Mark the client file to indicate that the account has been turned over to the collection agency and make no further attempts to collect it yourself. If the client contacts you about the account, refer him or her to the collection agency and do not enter into any discussion about the account or agree to accepting direct payments. Client files that have been turned over to a collection agency should be filed separately.

Payroll Accounting

Wage and salary records provide employers with information pertaining to a significant part of operating expenses. Accurate payroll records are necessary to calculate the federal, state, and local tax obligations of the business. Employers must also withhold a certain amount of taxes from employees' gross earnings to comply with tax laws. It is unlikely that members of the veterinary office team (employees) will become involved in the complicated paperwork and tax concerns of the practice but being aware of this aspect of the business may help you understand the reason why an employee must complete specific forms upon hire.

Form I-9

Before anyone can be officially hired, all employees must complete the **Employment Eligibility Verification Form**, or Form I-9, to ensure that they are authorized to work in the United States (Figure 9-2). You must submit two proofs of eligibility (such as a valid driver's license and a social security card) to your employer, who will then fill out the remainder of the form and file it accordingly.

Employee Compensation

The Fair Labor Standards Act, or Wages and Hours law, requires employers to maintain comprehensive records for each employee, including the employee's name, social security number, address, occupation (job title), rate of pay, basis for wage payment (per hour or salary), hours worked, wages at regular rate, overtime premium, additions to wages or deductions from wages, net payment, and date paid. Many veterinary offices use standard forms for individual employee payroll records but today, almost all payroll records are kept in a computer file.

OMB No. 1615-0047; Expires 08/31/12

Department of Homeland Security
U.S. Citizenship and Immigration Services

**Form I-9, Employment
Eligibility Verification**

Read instructions carefully before completing this form. The instructions must be available during completion of this form.

ANTI-DISCRIMINATION NOTICE: It is illegal to discriminate against work-authorized individuals. Employers **CANNOT** specify which document(s) they will accept from an employee. The refusal to hire an individual because the documents have a future expiration date may also constitute illegal discrimination.

Section 1. Employee Information and Verification *(To be completed and signed by employee at the time employment begins.)*

Print Name: Last	First	Middle Initial	Maiden Name

Address *(Street Name and Number)*		Apt. #	Date of Birth *(month/day/year)*

City	State	Zip Code	Social Security #

I am aware that federal law provides for imprisonment and/or fines for false statements or use of false documents in connection with the completion of this form.

I attest, under penalty of perjury, that I am (check one of the following):

☐ A citizen of the United States

☐ A noncitizen national of the United States (see instructions)

☐ A lawful permanent resident (Alien #) _____

☐ An alien authorized to work (Alien # or Admission #) _____
    until (expiration date, if applicable - *month/day/year*) _____

Employee's Signature	Date *(month/day/year)*

FIGURE 9-2 Form I-9 (Source: Department of Homeland Security, U.S. Citizenship and Immigration Services.)

Gross Pay. **Gross pay** is the total amount earned before payroll deductions. The simplest method of determining an employee's gross earnings is by defining how an employee is to be paid and to establish the rate of pay for the employee. A **salary** is a fixed amount of payment per month or year, regardless of the number of hours worked. The most frequently used method of determining gross earnings is by the *hourly rate*. Gross pay equals the amount earned per hour multiplied by the number of hours worked. The Fair Labor Standards Act sets minimum hourly wages. By law, a minimum rate of one and one-half times the regular hourly rate must be paid for all hours worked in excess of 40 hours per week. Higher rates, often twice the base rate, can also be paid as a premium for working at night, on Sundays or holidays. Overtime or premium pay requirements do not apply to the employees on a straight salary.

Net Pay. **Net pay** is the amount received after payroll deductions, which may include agreed-on deductions such as medical and dental insurance premiums, voluntary savings or investment plans. Social security contributions and income taxes, both state and federal, will also be deducted.

Taxes Withheld

Each week, a certain portion of your wages will be withheld from your check to go toward your income taxes. At the end of the year, this total amount withheld is reported on a W-2 form, which you submit with your income tax return. If too much money has been withheld throughout the year, you will receive a tax refund. If not enough has been withheld, you will need to send payment to make up the difference.

Federal Income Tax Withholding. The amount of federal tax income withheld from gross wages depends on the number of personal withholding allowances the employee has claimed. The most common of these withholding allowances are **exemptions** claimed for an employee and an employee's spouse and dependents. For each exemption claimed, a certain amount of an employee's yearly gross wages is exempt from income tax. An additional exemption can also be claimed for a dependent spouse who is 65 or older, or if the employee or dependent spouse is blind. To document the number of exemptions claimed, the employee must fill out an **Employee's Withholding Allowance Certificate**, better known as the Form W-4 (Figure 9-3).

The amount of federal income tax withheld from an employee's gross wages is determined by the **Federal Wage-Bracket Withholding Table**. Separate tables are available for weekly, biweekly, semimonthly, monthly, and miscellaneous pay periods for persons who are married and for persons who are unmarried. Using this table, the employer checks the number of exemptions claimed on employee's W-4 form to locate the employee's wage bracket to determine how much tax should be withheld. Almost all payroll and deductions are calculated by financial software in the computer program when an employee's details have been entered in the computer at the time of hire.

FICA Tax Withholding. In addition to income taxes, employers must withhold taxes levied under a law called the **Federal Insurance Contribution Act** (FICA)—social security taxes. These FICA taxes are imposed, in equal amounts, on *both* covered employers and their employees. The employer's share is called a **payroll tax**. Congress establishes the tax rate, and there is a cap on the amount of wages subject to the FICA tax. Social security benefits are based on

Form **W-4**
Department of the Treasury
Internal Revenue Service

Employee's Withholding Allowance Certificate

▶ **Whether you are entitled to claim a certain number of allowances or exemption from withholding is subject to review by the IRS. Your employer may be required to send a copy of this form to the IRS.**

OMB No. 1545-0074
2010

1 Type or print your first name and middle initial. | Last name | **2 Your social security number**

Home address (number and street or rural route)

3 ☐ Single ☐ Married ☐ Married, but withhold at higher Single rate.
Note. If married, but legally separated, or spouse is a nonresident alien, check the "Single" box.

City or town, state, and ZIP code

4 **If your last name differs from that shown on your social security card, check here. You must call 1-800-772-1213 for a replacement card.** ▶ ☐

5 Total number of allowances you are claiming (from line **H** above **or** from the applicable worksheet on page 2) 5
6 Additional amount, if any, you want withheld from each paycheck . . . . 6 $
7 I claim exemption from withholding for 2010, and I certify that I meet **both** of the following conditions for exemption.
- Last year I had a right to a refund of **all** federal income tax withheld because I had **no** tax liability **and**
- This year I expect a refund of **all** federal income tax withheld because I expect to have **no** tax liability.

If you meet both conditions, write "Exempt" here . . . . ▶ 7

Under penalties of perjury, I declare that I have examined this certificate and to the best of my knowledge and belief, it is true, correct, and complete.

Employee's signature
(Form is not valid unless you sign it.) ▶ **Date** ▶

8 Employer's name and address (Employer: Complete lines 8 and 10 only if sending to the IRS.) | 9 Office code (optional) | 10 Employer identification number (EIN)

For Privacy Act and Paperwork Reduction Act Notice, see page 2. Cat. No. 10220Q Form **W-4** (2010)

FIGURE 9-3 Employee's Withholding Allowance Certificate (Source: Department of the Treasury, Internal Revenue Service.)

the average earnings of the worker during the years of employment. To keep track of gross earnings and FICA taxes withheld, each employee must have a nine-digit **social security number** (application forms are available from local Social Security offices, Internal Revenue offices, and post offices).

State and Local Tax Withholding. In addition to federal income taxes and FICA taxes, employers in most states and some cities must also deduct state and city income taxes from their employees' earnings. The amount to be withheld is based on an employee's gross wages and, in some states, claimed exemptions.

Form W-2. At designated times during the year, the employer must pay all payroll taxes and employee withholdings to the proper agencies. Within one month after the end of a calendar year, the employer must furnish each employee with a **Wage and Tax Statement**, or Form W-2, which shows the amount of wages earned and the amounts of FICA, federal, state, and local income taxes that were withheld during the preceding year (Figure 9-4).

Form 941. The veterinarian, as an employer, must file an **Employer's Quarterly Federal Tax Return**. The employer must furnish information pertaining to total wages subject to withholding, the amount of federal income

22222 | **a** Employee's social security number 111-11-111 | OMB No. 1545-0008

b Employer identification number (EIN): 24-02060310

c Employer's name, address, and ZIP code:
Robert Barton, V.M.D.
4401 Birch Street
Newport Beach, CA 92663

d Control number

e Employee's first name and initial | Last name | Suff.
Mary H.Blatt
22 Elm Street
Newport Beach, CA 92663

f Employee's address and ZIP code

1 Wages, tips, other compensation: 10,400.00	**2** Federal income tax withheld: 821.60
3 Social security wages: 10,400.00	**4** Social security tax withheld: 795.60
5 Medicare wages and tips	**6** Medicare tax withheld
7 Social security tips	**8** Allocated tips
9 Advance EIC payment	**10** Dependent care benefits
11 Nonqualified plans	**12a** Code
13 Statutory employee ☐ Retirement plan ☐ Third-party sick pay ☐	**12b** Code
14 Other	**12c** Code
	12d Code

15 State	Employer's state ID number	**16** State wages, tips, etc.	**17** State income tax	**18** Local wages, tips, etc.	**19** Local income tax	**20** Locality name
CA		10,400.00	228.80	10,400.00	104.00	Newport Beach

Form **W-2** **Wage and Tax Statement** 2010 Department of the Treasury—Internal Revenue Service

Copy 1—For State, City, or Local Tax Department

FIGURE 9-4 Wage and Tax Statement, Form W-2 (Source: Department of the Treasury, Internal Revenue Service.)

tax withheld from the employees, the total wages subject to FICA taxes, and the combined amount of employees' and employer's FICA tax.

Unemployment Insurance

The Social Security Act of 1935 provided for temporary benefits for those who become unemployed as a result of economic conditions beyond their control. Unemployment insurance benefits are paid and administered by individual state departments of unemployment. States differ in regard to the types of covered employment and the number of workers an employer must have before the tax is levied. Federal funds are raised by a payroll tax on employers who qualify.

SUMMARY

Collecting payment from clients is an important part of veterinary office practices. You will need to be patient, respectful, and discreet when requesting payment. Oftentimes, you will also need to handle questions and complaints about fees, although the veterinarian sets the fees, not the veterinary staff.

When services are provided prior to payment, credit is being extended to the client for the duration of the office visit but require payment at the time services are rendered, that is, upon check out with the front desk. In some circumstance, the veterinarian may extend credit to a client for a longer period of time with a credit application and agreement of terms.

Most offices use computer accounting programs for billing and statements. These programs make it easier to locate, enter, and retrieve information.

Monthly statements should be sent to clients with a balance of one dollar or more, to remind them that payment is due. When a bill is 60 days overdue, it is considered past due, so you will need to send some reminders, either over the phone or through the mail, remembering to be respectful, but firm. Bills that are over 90 days past due are considered problem collections and may be turned over to a collection agency, which will take 20 percent to 50 percent of the amount collected. It is also possible to take the person to court, but that can be time-consuming and expensive. You need to be familiar with your office's own billing and collections policies, as they can differ from practice to practice.

SCENARIO

It was the 25th of the month again, and the receptionist was reviewing client accounts to identify any aging accounts receivable. He noticed that Mrs. P. still had not paid the remaining $100 balance owed for her poodle's surgery two months previously, so he mailed her a friendly, but firm, reminder of the amount due.

A week later, payment had not arrived, so he gave her a call. Unfortunately he got her answering machine, and left a message asking her to call back. She never did. Several follow-up calls garnered the same result.

A second statement marked "final notice" and warning that the invoice would be turned over for collection if not paid went out on the 25th of the next month, and more calls were made, but to no avail. The bill was finally turned over to the collection agency, and within two weeks, payment had been made (less 25 percent for their fee). A week later, Mrs. P. sent an angry letter to the office, complaining of harassment and indicating that she and her pet would not be back.

- Which fee collection guidelines did the receptionist follow?
- Was Mrs. P's complaint of harassment justified?
- What could have been done to prevent this collection problem?

REVIEW

For items 1–10, choose the correct term in each group.

1. Fees are determined by the (veterinary office assistant/veterinarian).
2. An encounter form is designed for (large fees/several purposes).
3. (Written reminders/telephone calls) least invade a customer's privacy in attempting to collect fees due to a veterinarian.
4. You (should/should not) attempt to provide cost estimates for customers who request them.
5. On credit applications you may not ask for the applicant's (date of birth/sex).
6. You must complete a federal Truth in Lending statement only for those customers who (are paying finance charges/have agreed to pay in a certain number of installments).
7. Adjusting a fee is unwise when the patient (has died/is a veterinarian).
8. What is cycle billing?
9. What information should you provide to a collection agency?

ONLINE RESOURCES

How to Collect Client Debt

This article explains how to collect on past due invoices while keeping clients, ranging from the gentle reminder to collection options, all the way to litigation. It requires the user to sign-in and is a part of the Score.org, part of the How-To on-line network.

http://www.score.org
Search Terms: Client debt; Workshops

How to Get Paid On Time

Also from Entrepeneur.com, this article provides suggestions and strategies for getting paid on time.

http://www.entrepreneur.com
Search Terms: Collecting; Payments; Your business

Glossary

A

account history The computerized version of a client account, containing client information and transaction history.

accounts receivable The total amount of payments outstanding or money owed by clients.

accounts receivable ratio The number of current accounts receivable compared to the average gross monthly charge.

admitted A patient who has been accepted into the hospital to receive treatment or a surgical procedure.

aging accounts receivable The process of determining the amount of time accounts outstanding have not been paid.

alpha search Searching a computer database by typing in the first few letters of the client's last name.

alphabetic filing A filing system based on the client's last name; each letter of the alphabet has its own color-coded stickers for easier retrieval and filing.

American Veterinary Medical Association (AVMA) The professional association dedicated to advancing the field of veterinary medicine and all of its related aspects.

anesthesia record sheet A minute by minute handwritten document recording the entire time an animal is anesthetized. Notations are made by the anesthetist regarding heart and respiration rates, percentage of anesthetic gas delivered, and additional medications and events which occur during the period of anesthesia.

B

body language The gestures and mannerisms that help people communicate.

broadband A high-speed Internet connection that allows another type of signal to travel over the same line simultaneously. Access is often provided through television cable companies. DSL service provided over the telephone line is also a form of broadband Internet access.

burnout A term used to describe a condition that results from too much stress over an extended period.

C

call board The component of the call director, containing buttons representing the different stations in the office.

call director A telephone unit with stations that allow several calls to come in at once.

capital equipment Equipment that has a fairly long life expectancy and contributes to the generation of practice revenues.

CD-R A CD-ROM that allows a user to write data to the disk with the use of a disk drive. A CD-R can only be written to once as a "read only" file and cannot be edited. Another type of CD-ROM, the CD-RW, allows the user to overwrite and/or edit the information repeatedly.

central processing unit The "brains" of the computer; the component of a working computer that houses the hard drive and other built-in components depending on the features of computer.

clean as you go The practice of putting things back where they belong after use, cleaning up an area after a procedure, clearing counters that have become unnecessarily cluttered, and maintaining a clean, clear workspace.

communicate Exchange or receive information.

communication process Relates to the four elements of communication: message, the information to be sent; the person sending the message; channel, the method through which the message is sent; and the receiver.

confidentiality Maintaining the privacy of client personal information, including financial status or a patient's medical records.

consent form A written legal document signed by the client stating that he or she gives his or her informed consent for the procedures to be performed and acknowledges the potential risks.

controlled substance logbook A written record of the use of controlled substances that contains the date, client's name, patient's name, the amount dispensed, the amount remaining, the name of veterinarian authorizing the use of the medication, and the individual dispensing the medication.

conventional format Another term for the source-oriented medical record which contains the patient's medical history in a chronological order.

cost estimate An approximation of fees, in writing, detailing the charges for proposed procedures and tests related to medical treatment. The cost estimate is provided before treatment is provided.

courtesy Being polite, showing good manners and respect

credit application A form filled out by a client seeking services with the understanding that the client will pay for them later. Applications inquire about employment history, current and past creditors, other forms of income, and references.

credit bureau An organization that collects credit information about individuals and provides reports to potential creditors, usually with a score or credit rating based on a person's previous repayment history.

cross-training Learning other people's duties to the level that you can perform their jobs if needed.

cycle billing A billing method in which accounts are grouped alphabetically and then divided into groups according to the number of statements that need to be sent. Each group of statements is sent out on a predetermined date each month.

D

database A collection of information organized into easily searchable computer files.

defense mechanisms Adjustments we make in our behavior, usually unconsciously, to help us deal with the experiences or feelings that cause psychological stress.

denial Refusing to admit or acknowledge that a traumatic, stressful situation exists in order to avoid dealing with it.

digital camera station The docking port for a digital camera; the device used to transfer data and photographs from the camera to a computer file.

discharge To formally release a patient from the care of a veterinary hospital.

displacement Transferring emotions from one situation to another, to make a difficult situation easier to accept and deal with.

dosimitry badge A badge worn by staff working in the radiology room. It absorbs scatter radiation so that the badges can be read by a dosimitry service to determine the amount of accumulated exposure to radiation for an individual over a set period of time.

E

empathy The ability to feel and understand what another person is feeling.

Employee's Withholding Allowance Certificate Form W-4; the form completed for the number of tax exemptions claimed by an employee.

Employer's Quarterly Federal Tax Return Form 941; a tax return that must be filed by the employer four times a year, stating the total amount of wages paid and the amount of taxes withheld.

Employment Eligibility Verification Form Form I-9; the form that must be filled out with the government to ensure that the person is authorized to work in the United States.

encounter form A billing document or travel sheet that contains basic client information and spaces to indicate procedures performed during the visit.

enunciation The way a person forms or articulates words.

establishing the matrix The advanced preparation of the appointment book.

ethics Rules of moral conduct.

euthanasia The practice of ending an animal's life humanely and painlessly.

exemptions Withholding allowances, generally for the employee's spouse, children, or other dependants.

extranet A computer network that allows limited access to off-site users. A user name and password is required to gain access.

F

Federal Insurance Contribution Act, FICA The law that requires contributions to be paid by both the employer and the employee for social security.

Federal Wage-Bracket Withholding Table The chart used by employers to determine the amount of federal tax that needs to be withheld and subtracted from an employee's paycheck.

fee profile A breakdown of average fees charged by veterinary practices and hospitals for similar services in a particular geographical area.

fee structure The way a practice charges for its services. The fee structure may be a flat fee per procedure, or it may be based on the amount of time spent with the client, or a combination of these methods.

feedback The return message in the communication process.

firewall A device or program used to limit access to a server.

fixed office hours A scheduling practice in which office hours are set, rather than making appointments. With the exception of emergencies, clients and patients are seen in the order of arrival.

flow scheduling Scheduling patients for segments, based on the amount of time an average appointment lasts.

footbath A shallow tray of water containing a disinfectant that people step into to avoid tracking potential pathogens from one area to another.

G

giardia A microscopic intestinal parasite which is easily transmitted from animal to animal, including humans.

gross pay The amount earned before payroll deductions.

H

hard drive The computer component that drives or manages an inserted disk.

hardware The physical components of a computer.

health certificate A certificate required for interstate and international transport of an animal which must be signed only by the veterinarian after a health exam.

I

induced The first period of anesthesia delivery.

inhalent Any substance that is breathed in, such as an anesthetic gas.

intellectualization Using reasoning to avoid the truth, as a way of denying strong feelings that may be socially unacceptable or difficult to accept.

Internet The global network of computer networks that allows millions of computer users to exchange information. There is no central server or single primary Internet source. Each individual server on the Internet can be used to send and receive information. Access to the Internet is obtained through an Internet service provider.

Internet service provider A company that provides Internet access through its servers. Most offer monthly subscriptions and require a user name and password that allow the subscriber to browse the World Wide Web and access e-mail providers.

interpersonal communication skills The art of communicating with other people.

intranet An internal computer network used in a single location. A user name and password is required to access the intranet, and a firewall is used to block unauthorized users.

inventory The products and supplies a veterinary practice has on hand.

J

jurisdiction The area in which a governing body has the legal power to administer and enforce the law.

K

keyboard An input device that allows the user to enter letters, numbers, and symbols into various computer programs.

kindness Being helpful, compassionate, and friendly, and treating others how you would want to be treated if the situation were reversed.

L

laboratory logbook Provides a synopsis of the laboratory tests requested and performed. It includes the date, client and patient names, tests performed in-house or sent to an outside laboratory.

listening Paying attention to what is being said, how it is being said, and the nonverbal messages (including body language) being sent.

logbooks Detailed written records on procedures relating to radiology, anesthesia/surgery, and controlled substances.

M

malingering Deliberately pretending to be sick in order to avoid work or a situation that causes anxiety.

master problem list A general health record indicating problems the patient is experiencing, as reported by the client or observed by a member of the veterinary team; contains dates of inoculation, results of all tests, drug allergies, and ongoing medical conditions.

material safety data sheets (MSDS) Published by the manufacturers of each product used by the practice, these sheets convey special handling requirements and user restrictions.

medical records The documentation that provides a detailed description of medical concerns, treatments/procedures, and progress notes. It is also a legal document that provides a complete and accurate history of each visit.

medical supplies Items used in providing medical care such as: bandaging and suture material, fluids, syringes, and needles. In short, everything required to administer medical care.

message The information sent or received in communication.

modem A device that allows the computer user to connect to another computer, network, or the Internet.

monitor The computer screen.

monotone Not showing a change in feeling or pitch while speaking.

mouse The device used to manipulate the cursor or pointer on the screen.

N

National Association of Veterinary Technicians in America (NAVTA) An organization of veterinary technicians dedicated to fostering high standards of veterinary care, promoting the health-care team, and advancing the veterinary technology profession.

National Institute for Occupational Safety and Health (NIOSH) The federal agency responsible for conducting research and making recommendations for the prevention of work-related disease and injury.

net pay The amount of earnings received after payroll deductions.

network Two or more computers connected to one another.

neuter or spay certificate   A certificate required to prove that an animal is no longer sexually intact.

numerical filing   A filing system based on each client's or patient's assigned number, rather than the name.

O

objective information   Factual, measurable data, such as an animal's weight.

Occupational Safety and Health Administration (OSHA) Provides guidelines created by the U.S. Department of Labor for safe and healthy working conditions.

operating system   The program that runs the computer and all of its other software applications. The most common operating systems are Microsoft Windows and Mac OS.

P

paraphrase   To use different words to express the same idea.

pathogen   Any disease causing organism.

patience   The ability to bear trials calmly without complaint.

payroll tax   The employer's share of FICA taxes.

personnel manual   Document that outlines the employer's expectations and provides a basis of consistency in how practice issues are handled.

pharmaceuticals   Medicines including vaccines, antibiotics, sedatives, ointments, and other dispensed drugs.

pitch   The highness or lowness of sound in a person's voice.

practice manager   A person in a veterinary practice responsible for overseeing and coordinating the behind-the-scene tasks and duties that allow the efficient delivery of medical care to patients and the accommodation of clients.

prejudice   Preconceived biases and opinions.

printer   The equipment that prints out text or images from the computer.

prioritize   To list tasks by importance, with the most important at the top and the least important at the bottom.

problem-oriented medical record   Organizing a medical record in detail for each patient and the problem or concern that is presented.

procedures manual   Information detailing the practice policies, a "how to" manual for each different aspect of the practice.

professional courtesy   Providing professional services to another professional person or veterinarian at a lower fee.

projection   One's own ideas, feelings, or attitudes are attributed to someone or some thing else.

psychosomatic illness   Real physical symptoms resulting from an emotional or mental condition.

purge   To remove client and patient files that are no longer active. These must be retained in a separate area for a period of at least seven years.

R

radiology   Refers to the use of X-rays.

rationalization   Attributing one's actions to logical reasons.

receptionist   The person in a practice who is responsible for answering the phone, setting appointments, greeting clients as they enter, and any other front desk duties.

reference points   Factors that determine how messages are expressed and understood; may include education, experience, social and cultural barriers, and religious beliefs.

regression   Returning to an earlier mental or behavioral state during times of high stress.

relate   The facilitation of communication by having something in common with the other person.

reminder cards   Postcards that are sent to clients reminding them of vaccination due.

repression   Socially unacceptable or painful desires or impulses are pushed out of the conscious mind into the unconscious mind.

right-to-know station   Houses the practice's safety information.

S

safety manual   Policies and instructions about minimizing the risks of accidents and injuries.

salary   The money earned by an employee on a weekly, monthly, or annual basis.

scanner   An imaging device used to make a digital image of a document or photograph, so that it can be transferred to a computer file for storage or transmission.

screening   Handling or directing incoming calls according to the nature of the call.

server   A computer or device that manages network resources.

sharps container   A heavy plastic container with a lid that can be permanently sealed, used for the safe disposal of hypodermic needles and scalpel blades.

signalment   Details of the patient's breed, color, sex, age, and reproductive status.

slander   A verbal statement that harms a person's image or reputation, either intentionally or unintentionally.

SOAP   Progress notes in a problem-oriented medical record divided into Subjective, Objective, Assessment, and Procedure sections.

social security number   The number assigned to American citizens by the Social Security Administration. All employees must have a social security number.

software Computer programs.

source-oriented medical record Organizing medical records in each patient's file in chronological order.

station Individual unit of a call director telephone system, in which a call can be taken while other calls are coming in simultaneously.

statute of limitations The amount of time in which a practice can take legal action against a client for nonpayment.

stereotype Preconceived ideas about a group of people made without taking individual difference into account.

stress The physical and/or psychological changes that occur in the body as a result of a change and or continued mental pressure in the environment.

subjective information Nonmeasurable data that describes an animal's attitude.

sublimation Diverting an instinctual desire or impulse into a socially acceptable activity.

substance abuse The overuse, inappropriate use, or illegal use of drugs, alcohol, or other potentially harmful substance.

T

tact Doing and saying the right things at the right time.

telephone reminder A telephone call to a client to remind him or her of a balance due on an account.

temporary withdrawal Finding ways to avoid dealing with a painful or difficult situation.

thumb drive see USB Flash drive.

tickler file A file similar to an appointment book, with slots for mail and telephone messages, organized by days of the month.

time audit A method of time management that involves noting all the tasks performed throughout the day and how much time is spent on each task. This helps to recognize time use patterns, which helps to develop strategies for more efficient use of time available…

time tracking Keeping track of time as you work, in order to prevent yourself from spending too much time doing something to the exclusion of something else.

tone Vocal quality that expresses mood or feeling.

toner A special kind of dry, powdered ink used in laser printers and photocopiers. Toner generally comes in long cartridges specifically designed for individual printers or copiers.

toxoplasmosis A zoonotic coccidian parasite in the intestine of all cats, including domestic pets.

travel sheet A preprinted form attached to a patient's file used to record each aspect of care received and to be billed.

U

USB flash drive A small (approximately 2 inch) data storage device used to download and transport files from computer to computer. Also called a thumb drive.

V

vaccination certificate The document that states the vaccination status of an animal.

verbatim Word for word.

veterinarian An individual licensed to practice veterinary medicine; the member of the veterinary health team ultimately responsible for diagnosing, prescribing treatments, and performing surgery.

veterinary assistant The person responsible for assisting the veterinary technician and veterinarian as needed; duties range from providing a limited degree of medical care to the clerical duties of the front desk and reception.

veterinarian–client–patient relationship (VCPR) The relationship between the veterinarian and the client and patient.

veterinary technician A trained professional who directly assists the veterinarian in the administration of health care to patients.

volume Degree of loudness.

W

Wage and Tax Statement Form W-2; a statement of the total amount earned by an employee and amount withheld for taxes.

want list Place to make notes of medications or supplies that have dropped to low quantities.

wave scheduling Scheduling the number of patients that can be seen in an hour's worth of segments.

Web site A virtual location on the World Wide Web.

word processing program Software that replaces the use of a typewriter, and includes additional features such as spell check, automatic formatting, and the ability to cut and paste text and images.

workstation An individual computer on a network.

written reminder Statement or a letter to a client to remind him or her of a balance due on an account.

X

X-ray logbook A written record of each radiograph that is taken. It includes patient information, views taken, and settings used to produce the X-ray.

Z

zoonotic, zoonose A disease that is transmitted from animals to humans.

Bibliography

Ballard, Bonnie and Rockett, Jody, *Restraint and Handling for Veterinary Technicians and Assistants*, Delmar, Cengage Learning, 2009.

Blood, D.C. Stoddert and Saunders, V.P., *Comprehensive Veterinary Dictionary*, 2nd ed., W.B Saunders, 1999.

Judah, Vicki and Nuttall, Kathy, *Exotic Animal Care & Management*, Delmar, Cengage Learning, 2008.

McCurrin, Dennis M. and Bassert, Joanna M., *Clinical Textbook for Veterinary Technicians*, 2nd ed., Mosby, 2002.

Sirois, Margi, *Principles and Practice of Veterinary Technology*, 2nd ed., Mosby, 2004.

http://www.accountingfordummies.com (accessed May 14, 2010)

http://www.avma.org (accessed February 18, 2010)

http://www.aplb.org (accessed January 14, 2010)

http://www.LawDepot.com (accessed February 18, 2010)

http://www.navta.org (accessed November 14, 2009; July 12, 2010)

http://www.osha.gov (accessed January 31, 2010)

http://www.veterinarytechnician.com (accessed November 16, 2009)

http://www.vhma.org (accessed August 30, 2009)

Index